Sara Yasemin Demiroglu

Apoptosis induced by cytotoxic cells of the immune system

Sara Yasemin Demiroglu

Apoptosis induced by cytotoxic cells of the immune system

The role of heat shock protein 70 and sulphatases 1 and 2 in granzyme B-induced apoptosis

Südwestdeutscher Verlag für Hochschulschriften

Impressum / Imprint
Bibliografische Information der Deutschen Nationalbibliothek: Die Deutsche Nationalbibliothek verzeichnet diese Publikation in der Deutschen Nationalbibliografie; detaillierte bibliografische Daten sind im Internet über http://dnb.d-nb.de abrufbar.
Alle in diesem Buch genannten Marken und Produktnamen unterliegen warenzeichen-, marken- oder patentrechtlichem Schutz bzw. sind Warenzeichen oder eingetragene Warenzeichen der jeweiligen Inhaber. Die Wiedergabe von Marken, Produktnamen, Gebrauchsnamen, Handelsnamen, Warenbezeichnungen u.s.w. in diesem Werk berechtigt auch ohne besondere Kennzeichnung nicht zu der Annahme, dass solche Namen im Sinne der Warenzeichen- und Markenschutzgesetzgebung als frei zu betrachten wären und daher von jedermann benutzt werden dürften.

Bibliographic information published by the Deutsche Nationalbibliothek: The Deutsche Nationalbibliothek lists this publication in the Deutsche Nationalbibliografie; detailed bibliographic data are available in the Internet at http://dnb.d-nb.de.
Any brand names and product names mentioned in this book are subject to trademark, brand or patent protection and are trademarks or registered trademarks of their respective holders. The use of brand names, product names, common names, trade names, product descriptions etc. even without a particular marking in this work is in no way to be construed to mean that such names may be regarded as unrestricted in respect of trademark and brand protection legislation and could thus be used by anyone.

Verlag / Publisher:
Südwestdeutscher Verlag für Hochschulschriften
ist ein Imprint der / is a trademark of
OmniScriptum GmbH & Co. KG
Heinrich-Böcking-Str. 6-8, 66121 Saarbrücken, Deutschland / Germany
Email: info@svh-verlag.de

Herstellung: siehe letzte Seite /
Printed at: see last page
ISBN: 978-3-8381-1170-4

Zugl. / Approved by: Göttingen, Georg-August-University, Diss., 2009

Copyright © 2009 OmniScriptum GmbH & Co. KG
Alle Rechte vorbehalten. / All rights reserved. Saarbrücken 2009

Contents

1 List of abbreviations   **1**
- 1.1 Amino acid one-letter code 1
- 1.2 Abbreviations 1

2 Introduction   **7**
- 2.1 Apoptosis 7
- 2.2 Induction of apoptosis by cytotoxic cells of the immune system 9
- 2.3 Role of *Sulf1* and *Sulf2* on heparan sulphates and in the uptake of granzyme B into target cells 14
- 2.4 The role of heat shock protein 70 in apoptosis 15
  - 2.4.1 Heat shock protein 70 15
  - 2.4.2 Heat shock protein 70 in apoptosis 16
- 2.5 Aims 19

3 Materials   **21**
- 3.1 Antibodies and dyes 21
- 3.2 Primers 23
- 3.3 Chemicals and reagents 24
- 3.4 Kits 27
- 3.5 Buffers and stock solutions 27
- 3.6 Cell lines, viruses and media 31
- 3.7 Laboratory animals 33
- 3.8 Used laboratory equipment 34
- 3.9 Disposable plastic ware and other disposables 36
- 3.10 Computational analysis 36
- 3.11 List of providers 37

4 Methods   **41**
- 4.1 Cell culture methods 41
  - 4.1.1 Culturing of cells 41

Contents

- 4.1.2 Subcloning of the Wt clone of *Sulf* mouse embryonic fibroblasts by limiting dilution 41
- 4.1.3 Acute induction of heat shock protein 70 in the human melanoma cell line Ge-tet 42
- 4.1.4 Induction of apoptosis using staurosporine 42
- 4.1.5 Optimising multiplicity of infection for adenovirus 42
- 4.1.6 Induction of apoptosis using granzyme B and adenovirus 43
- 4.1.7 Uptake of labelled granzyme B into cells with acute heat shock protein 70 overexpression and mouse embryonic fibroblasts 43
- 4.2 Immunological methods 43
  - 4.2.1 Generation of effector cells 43
    - 4.2.1.1 Generation of SIINFEKL-specific cytotoxic T-lymphocytes from transgenic OT-I mice 44
    - 4.2.1.2 Generation of human natural killer cells from whole blood by density gradient centrifugation and negative MACS selection .. 44
  - 4.2.2 Preparation of concanavalin A supernatants for the stimulation of effector cells 45
  - 4.2.3 Cytotoxic assays 45
    - 4.2.3.1 ^{51}Chromium release assay 45
    - 4.2.3.2 [^{3}H]-Thymidine release assay 46
  - 4.2.4 Flow cytometric analyses 47
    - 4.2.4.1 Cell surface stainings for flow cytometric analysis 47
    - 4.2.4.2 Intracellular flow cytometric analysis 47
    - 4.2.4.3 DiD-staining of Ge cells for activation of caspase-3 after NK cell-induced apoptosis 48
  - 4.2.5 Measurement of apoptosis in cells 48
    - 4.2.5.1 Annexin V binding to phosphatidylserine on the cell surface .. 49
    - 4.2.5.2 Release of cytochrome c from mitochondria 49
    - 4.2.5.3 Change in mitochondrial membrane potential 49
    - 4.2.5.4 Activation of caspase-8 50
    - 4.2.5.5 Activation of caspase-3 50
    - 4.2.5.6 Sub G1-peak analysis to measure DNA loss 50
    - 4.2.5.7 Apoptotic ladder to measure DNA fragmentation 50
- 4.3 Biochemical methods 51
  - 4.3.1 Preparation of cell lysates for immunoblot analysis 51
  - 4.3.2 SDS-PAGE 51
  - 4.3.3 Immunoblot 52

Contents

	4.3.4	Densitometric analysis of levels of coxsackie and adenovirus receptor on *Sulf* mouse embryonic fibroblasts	53
4.4	Molecular biological methods		53
	4.4.1	RNA isolation from cells	53
		4.4.1.1 RNA isolation	54
		4.4.1.2 Determining RNA concentration	54
		4.4.1.3 Determining RNA quality by RNA 6000 Pico Chip analysis	55
	4.4.2	Microarray analysis	56
	4.4.3	Quantitative real-time PCR	60
		4.4.3.1 Transcription of RNA into cDNA with reverse-transcriptase PCR	60
		4.4.3.2 Validating primers for quantitative real-time PCR with standard PCR	61
		4.4.3.3 Agarose gel electrophoresis	62
		4.4.3.4 Quantitative real-time PCR using SYBR green	62
		4.4.3.5 Evaluation of data using Pfaffl	63
	4.4.4	Comparison of microarray and quantitative real-time PCR expression data	64

5 Results 67

5.1	Role of heat shock protein 70 in apopotosis		67
	5.1.1	HSP70 overexpression in different Ge clones	67
	5.1.2	Gene expression analysis of cells acutely overexpressing heat shock protein 70	67
		5.1.2.1 Whole human genome microarray analysis of Ge-tra and Ge-tet-1 cells	67
		5.1.2.2 Quantitative real-time PCR analysis of selected genes	74
	5.1.3	Effect of acute and permanent overexpression of heat shock protein 70 on early and late apoptosis	75
		5.1.3.1 Effect of acute HSP70 overexpression on phosphatidylserine exposure after granzyme B-induced apoptosis	75
		5.1.3.2 Effect of acute HSP70 overexpression on granzyme B-induced DNA fragmentation	77
		5.1.3.3 Effect of permanent HSP70 overexpression on granzyme B-induced DNA fragmentation	79
		5.1.3.4 Effect of acute HSP70 overexpression on granzyme B uptake	80
		5.1.3.5 Effect of acute HSP70 overexpression on phosphatidylserine exposure after staurosporine-induced apoptosis	80

| | | 5.1.3.6 | Effect of acute HSP70 overexpression on staurosporine-induced DNA fragmentation . . . . | 83 |
| | | 5.1.3.7 | Effect of permanent HSP70 overexpression on staurosporine-induced DNA fragmentation . . . . | 83 |

- 5.1.4 Influence of acute HSP70 overexpression on key steps in apoptosis 83
  - 5.1.4.1 Effect of acute HSP70 overexpression on the change in mitochondrial membrane potential $\Delta\Psi$ 85
  - 5.1.4.2 Effect of acute HSP70 overexpression on release of cytochrome c from mitochondria 86
  - 5.1.4.3 Effect of acute HSP70 overexpression on activation of initiator caspase-8 90
  - 5.1.4.4 Effect of acute HSP70 overexpression on activation of effector caspase-3 91
  - 5.1.4.5 Effect of acute HSP70 overexpression on DNA fragmentation analysed by apoptotic ladder 96
- 5.2 Role of sulphatases 1 and 2 in apoptosis 98
  - 5.2.1 Uptake of granzyme B into *Sulf* mouse embryonic fibroblasts 99
  - 5.2.2 Effect of the deficiency of *Sulf1* and *Sulf2* in target cells on lysis by cytotoxic T-lymphocytes 99
  - 5.2.3 Effect of the deficiency of *Sulf1* and *Sulf2* in target cells on apoptotic killing by cytotoxic T-lymphocytes 100
  - 5.2.4 H2K^b expression levels on *Sulf* mouse embryonic fibroblasts 104
  - 5.2.5 Effect of the deficiency of *Sulf1* and *Sulf2* on apoptosis induced by granzyme B 108
  - 5.2.6 Effect of the deficiency of *Sulf1* and *Sulf2* on transfection efficiency of adenovirus type 5 109
  - 5.2.7 Expression of coxsackie and adenovirus receptor and integrin α_v on the cell surface of *Sulf* mouse embryonic fibroblasts 110
  - 5.2.8 Effect of heparinase II and III treatment of *Sulf* MEFs on adenoviral GFP expression 114
  - 5.2.9 Heparan sulphates as receptors for type 5 adenovirus 115

6 Discussion 117
- 6.1 Role of HSP70 in apoptosis 117
  - 6.1.1 Gene expression analysis of cells acutely overexpressing HSP70 117
  - 6.1.2 Effect of acute and permanent overexpression of HSP70 on early and late stages of apoptosis 119

		6.1.3 Analysis of key steps in apoptosis after the acute overexpression of HSP70	121
6.2	Role of sulphatases 1 and 2 in apoptosis		126
	6.2.1	Interaction of granzyme B and mouse embryonic fibroblasts deficient for *Sulf1* and *Sulf2*	127
	6.2.2	Uptake of adenovirus type 5 into mouse embryonic fibroblasts deficient for *Sulf1* and *Sulf2*	129

Summary	**137**
Bibliography	**139**
List of Figures	**165**
List of Tables	**169**
Acknowledgements	**171**

Appendix 172

A Quantitative real-time PCR	**173**
A.1 Dissociation curves	173
B Microarray	**175**
B.1 Quantification of labelled amplified cRNA for microarray	175
B.2 All genes found to be regulated in micoarray analysis	176

1 List of abbreviations

1.1 Amino acid one-letter code

A	Alanine
C	Cysteine
D	Aspartic acid
E	Glutamic acid
F	Phenylalanine
G	Gylcine
H	Histidine
I	Isoleucine
K	Lysine
L	Leucine
N	Asparaginine
P	Proline
Q	Glutamine
R	Arginine
S	Serine
T	Threonine
V	Valine
W	Trypthophane
Y	Tyrosine

1.2 Abbreviations

3-APA	3-aminophthalate

1 List of abbreviations

aa	amino acid
AdV	adenovirus
AIF	apoptosis inducing factor
ANOVA	analysis of variances
APAF1	apoptotic peptidase activating factor-1
APC	antigen presenting cell
APS	ammonium persulphate
ASNS	asparagine synthetase
ATP	adenosine triphosphate
BAD	BCL-2 antagonist of cell death
BAK	BCL-2-antagonist/killer-1
BAX	BCL-2-associated X protein
BCL-2	B-cell lymphoma-2
BCL10	B-cell lymphoma-10
BCL-XL	BCL-2-like protein
BH	BCL-2 homology
bp	base pairs
BID	BH3-interacting domain death agonist
BIK	BCL-2-interacting killer
BIM	BCL-2-like-11
BMF	BCL-2 modifying factor
BSA	bovine serum albumine
C	Celsius
$CaCl_2$	calcium chloride
CAD	caspase-activated DNase
CAR	coxsackie and adenovirus receptor
CASP8	caspase-8
caspase	cysteine aspartic acid-specific protease
CD	cluster of differentiation

1.2 Abbreviations

CHO	chinese hamster ovary
Con A	concanavalin A
CMV	cytomegalovirus
cpm	counts per minute
ct	cycle treshold
CTL	cytotoxic T-lymphocyte
DAB	diaminobenzoide
DFF45	DNA fragmentation factor, 45 kDa; also known as ICAD
Dko	double knock-out
DMEM	Dulbecco's modified Eagle's medium
DMSO	dimethylsulfoxide
DNA	desoxyribonucleic acid
DTT	dithiotreitol
E	efficiency
ECL	enhanced chemoluminescence
EDTA	ethylendiamine tetraacetic acid
EGTA	ethylene glycol tetraacetic acid
EIF5	eukaryotic translation initiation factor 5
EtOH	ethanol
FACS	fluorescence activated cell sorting
FasL	Fas ligand
FCS	fetcal calf serum
FITC	fluoresceinisothiocyanate
GAAD	GrA-activated DNase
GFP	green fluorescence protein
Gr	granzyme
GRB10	growth factor eceptor-bound protein 10
GRP	glucose regulated protein
H_2O	water

1 List of abbreviations

HCl	hydrochloric acid
Hepes	4-(2-hydroxyethyl)-1-piperazineethanesulfonic acid
HIV-1	human immunodeficiency virus-1
hr	hour
HRK	harakiri
HRP	horseradish peroxidase
HS	heparan sulphate
HSC70	heat shock cognate 70
HSP70	heat shock protein 70
HSF	heat shock factor
ICAD	inhibitor of CAD
ICAM-1	intercellular cell adhesion molecule-1
IFN	interferon
IL-2	interleukin-2
JC-1	5,5',6,6'-tetrachloro-1,1',3,3'-tetraethylbenzimidazolylcarbocyanine iodide
JNK	c-Jun-N-terminal kinase
JUND	jun D proto-oncogene
kb	kilo base pairs
kDa	kilo Dalton
LFA-1	leukocyte function-associated antigen-1
LOX-1	lectin-like oxidised low-density lipoprotein receptor-1
μCi	micro Curie
MACS	magnetic cell sorting
MCL1	myeloid cell leukaemia sequence-1
MEF	mouse embryonic fibroblast
MFI	mean fluorescence intensity
$MgCl_2$	magnesium chloride
MHC	major histocompatibility complex
MIC	MHC class I chain-related molecule

1.2 Abbreviations

min	minutes
MOI	multiplicity of infection
MPR300	mannose-6-phosphate receptor
MTOC	microtubule-organising centre
$NaHCO_3$	sodium bicarbonate
NaCl	sodium chloride
NaOH	sodium hydroxide
NFAT	nuclear factor of activated T-cells
NH_4Cl	ammonium chloride
NK	natural killer
NOXA	phorbol-12-myristate-13-acetate-induced protein 1
PAGE	polyacrylamide gel electrophoresis
PBMC	peripheral blood mononuclear cell
PBS	phosphate-buffered saline
PCR	polymerase chain reaction
PEG	polyethylene glycol
PFA	paraformaldehyde
pfu	plaque forming unit
PG	proteoglycan
PI	propidium iodide
$KHCO_3$	potassium bicarbonate
PS	phosphatidylserine
PUMA	BCL-2 binding component-3
qRT-PCR	quantitative real-time PCR
RIN	RNA integrity number
RNA	ribonucleic acid
RPE	red phycoerythrin
rpm	rounds per minute
RT	room temperature

1 List of abbreviations

SD	standard deviation
SEM	standard error of the mean
SDS	sodium dodecyl sulphate
SSPE	sodium chloride, sodium hydrogen phosphate, and EDTA
STC2	stanniocalcin 2
Sulf	sulphatase
TAE	Tris/acetate/EDTA
TBE	Tris/borate/EDTA
TBL1XR1	transducin β-like 1 X-linked receptor 1
TC	tri-colour
TCR	T-cell receptor
TdT	terminal desoxynucleotidyltransferase
TEMED	N,N,N,N-tetramethyl-ethane-1,2-diamine
THOC4	THO complex 4
TLR	Toll-like receptor
TM4SF1	transmembrane 4 L six family member 1
TNF	tumour necrosis factor
Tris	Tris(hydroxymethyl)-aminomethane
TXNRD1	thioredoxin reductase 1
UTR	untranslated region
UV	ultraviolet
Wt	wild type
x g	accelaration of gravity
XIAP	x-linked inhibitor of apoptosis protein

2 Introduction

2.1 Apoptosis

The word apoptosis is derived from the greek word απόπτωσις - apo meaning from and ptosis meaning falling, like leaves fall from trees in autumn. It is a form of programmed cell death (Degterev and Yuan 2008) which can be induced by milder insults to the cell in contrast to necrosis taking place after intense insult to the cell (McConkey 1998). In apoptotic cell death, a cell dyes in a controlled manner and damage and disruption to neighbouring cells in minimised (Kerr et al. 1972). Phagocytes finally remove the resulting cell debris without leakage of the cytoplasmic content of the cell into the surrounding tissue, which normally does not attract inflammatory immune cells to this site (Savill and Fadok 2000). Necrotic cell death in contrast evokes inflammation by activation of dendritic cells (Gallucci et al. 1999), macrophages and neutrophils and other cells of the innate immune response (Chen et al. 2007; Oppenheim and Yang 2005) but also cytotoxic T-lymphocytes (CTLs) (Shi et al. 2000). Apoptotic cell death shows distinct morphological features such as condensation and fragmentation of the nucleus, membrane blebbing and the formation of apoptotic bodies (Clarke 1990; Kerr et al. 1972; Wyllie et al. 1980). Apoptosis is an important cellular process during development (Twomey and Mc-Carthy 2005), as for example mice deficient for apoptotic peptidase activating factor-1 (*Apaf1*), a component of the apoptosome, show reduced apoptosis in the brain and hyperproliferation of neuronal cells (Yoshida et al. 1998). Apoptosis is also indispensable especially for the immune system as cytotoxic cells, CTLs and natural killer (NK) cells, can induce apoptosis in tumour or virus-infected cells to destroy them (Cohen et al. 1992). Apoptosis is also important for cell homeostasis in the immune system to shut-down an immune response after successful elimination of pathogens or cancer cells (Strasser and Pellegrini 2004). It is therefore not surprising, that a dysregulation of apoptosis causes a wide variety of diseases particularly cancer (Debatin et al. 2003; Fulda and Debatin 2004).

Main effectors of the signalling pathways during apoptosis are cysteine aspartic acid-specific proteases (caspases) and members of the B-cell lymphoma-2 (BCL-2) family.

Humans possess 11 caspases, which can be subdivided into initiator caspases (caspases-2, -8, -9, and -10), effector caspases (caspases-3, -6, and -7), and inflammatory caspases (caspases-1, -4, -5, and -12). Initiator caspases mainly activate effector caspases, which proteolytically

2 Introduction

cleave many substrates during apoptosis. Inflammatory caspases are activated during innate immune responses and are involved in the regulation of cytokine processing. Among the about 400 caspase substrates identified (Lüthi and Martin 2007), are many essential proteins of cell homeostasis and structure. Caspase substrates in the cytoskeleton are actin and tubulins and together with the loss of other components of the cytoskeleton, their loss causes the characteristic blebbing of the cells (Coleman et al. 2001; Cotter et al. 1992). The proteolysis of lamins A, B, and C, components of the nuclear lamina, by caspases causes a loss of laminar integrity and is one of the events leading to nuclear fragmentation (Rao et al. 1996). Caspase-3 can directly cleave inhibitor of CAD (ICAD) so that caspase-activated DNase (CAD) can cause chromatin fragmentation in the nucleus (Enari et al. 1998; Sakahira et al. 1998). Proteins involved in transcription (such as nuclear factor of activated T-cells (NFAT)c1 and NFATc2 and NFκBp65) and translation (translation initiation factors eIF2a, eIF3, eIF4B, eIF4E, eIF4G, and eIF4H) are also targets of caspases (Lüthi and Martin 2007) as well as ribosomal ribonucleic acid (RNA) (Houge et al. 1993). The cytoskeleton, the nucleus, and transcription and translation factors are targeted by caspases during apoptosis to destabilise the cell at many critical points. Only the fragmentation of mitochondria (Taylor et al. 2008) does not directly involve caspases. Caspases are only responsible for the cleavage of the p75 subunit of complex I of the electron transport chain, which is required for swelling and destructive morphological changes in mitochondria during apoptosis (Ricci et al. 2004). The main role in fragmentation of mitochondria has to be attributed to BCL-2 family members.

The family of BCL-2 proteins can be further subdivided according to their function and structure. They contain between one and four BCL-2 homology (BH) domains (BH1-4) (Youle and Strasser 2008). The members of the of anti-apoptotic BCL-2 subfamily contain 4 BH domains and most of them also encompass a transmembrane domain locating them to the membrane (such as BCL-2, BCL-2-like protein (BCL-XL), and myeloid cell leukaemia sequence-1 (MCL1)). The most prominent members of the pro-apoptotic multidomain BCL-2 subfamily are BCL-2-antagonist/killer-1 (BAK) and BCL-2-associated X protein (BAX), which synergistically form pores into the mitochondrial outer membrane (Kuwana et al. 2005; Letai et al. 2002). Members of this family lack the BH4 domain and interestingly, the loss of the BH4 domain due to cleavage by caspases converts anti-apoptotic BCL-2 members into pro-apoptotic ones (Cheng et al. 1997). The third BCL-2 subfamily comprises eight members (BCL-2 antagonist of cell death (BAD), BH3-interacting domain death agonist (BID), BCL-2-interacting killer (BIK), BCL-2-like-11 (BIM), BCL-2 modifying factor (BMF), harakiri (HRK), phorbol-12-myristate-13-acetate-induced protein 1 (NOXA), and BCL-2 binding component-3 (PUMA)), which only share a structural similarity in the BH3 domain and are therefore termed BH3-only proteins. When they are overexpressed they all promote apoptosis (Fukazawa et al. 2003; Hsieh et al. 2003; Huang and Strasser 2000; Liu et al. 2007; Willis and Adams 2005).

Anti-apoptotic BCL-2 family members inhibit the BH3-only protein-induced oligomerisation of BAX and BAK and with this prevent the release of cytochrome c and other mitochondrial intermembrane space proteins. Thus, anti-apoptotic BCL-2 family members prevent apoptosis by protecting mitochondrial integrity by direct interaction with pro-apoptotic BCL-2 family members and not by inhibiting caspase activity (Youle and Strasser 2008). On the other hand, apoptosis can be induced, when pro-apoptotic BH3-only proteins bind anti-apoptotic proteins, such as binding of BAD to BCL-2 and BCL-XL, or binding of NOXA to MCL1 (Chen et al. 2005). Apoptosis can also be induced, when BAX and BAK are activated by BH3-only proteins (Letai et al. 2002).

The various signalling pathways taking place during apoptosis are illustrated figure 2.1 on the following page and were recently reviewed (Taylor et al. 2008). Mainly two apoptotic pathways exist, an extrinsic pathway involving death receptors and caspases and an intrinsic pathway involving mitochondria but cross-talk takes place at many points (Barnhart et al. 2003; Zimmermann and Green 2001). The following description is just a simplified version of both pathways and their key molecules. As figure 2.1 on the next page illustrates, many more proteins are involved, inhibiting or activating certain other proteins.

The extrinsic pathway can be triggered by engagement of death receptors such as Fas receptor or of TNF receptor 1. This engagement leads to the activation of caspase-8, which can directly activate caspase-3 bypassing the mitochondria and finally leading to cell death. An interconnection between both pathways exists via caspase-8 through the cleavage of BID, which then blocks anti-apoptotic BCL-2 family members, releasing BAX and BAK (Li et al. 1998).

The intrinsic pathway can be triggered by factors such as growth factor depriviation, ultraviolet (UV)-radiation, DNA damage, stress but also viruses. These triggers activate BH3-only proteins. They can inhibit the anti-apoptotic BCL-2 family members by direct interaction (Willis et al. 2007), so that the two pro-apoptotic BCL-2 members BAX and BAK are no longer inhibited by anti-apoptotic BCL-2 family members. Thus, BAX and BAK can cause mitochondrial outer membrane depolarisation. Some BH3-only proteins might also directly activate BAX or BAK (Youle 2007). The released cytochrome c from mitochondria together with APAF1 and caspase-9 forms an assembly named apoptosome, which can activate the effector caspase-3, which cleaves several substrates leading ultimately to cell death.

2.2 Induction of apoptosis by cytotoxic cells of the immune system

Cytotoxic cells of the immune system to which NK cells and CTLs belong can destroy e.g., virus-infected cells and tumour cells. Antigen-specific activated CTLs possess at least two

2 Introduction

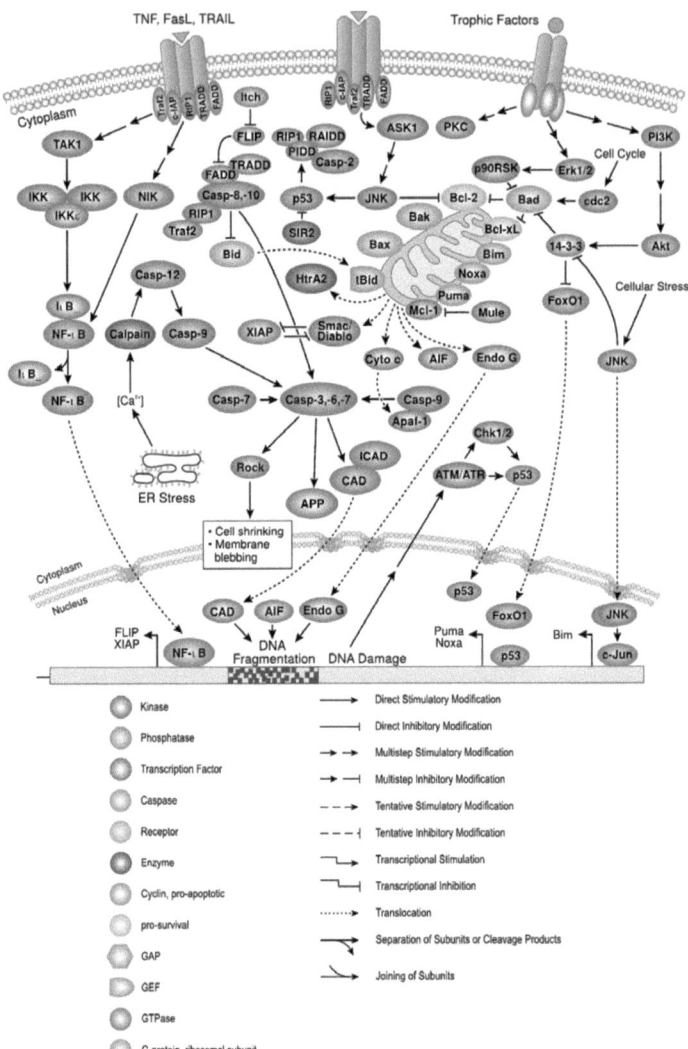

Figure 2.1: Signalling in apoptosis Apoptosis is a controlled cell death mechanism characterised by nuclear condensation, cell shrinkage, membrane blebbing and desoxyribonucleic acid (DNA) fragmentation. Caspases, a family of cysteine proteases, are the central regulators of apoptosis. Initiator caspases (including caspase-2, -8, -9, and -10) are closely coupled to pro-apoptotic signals. Once activated, these caspases cleave and activate downstream effector caspases (including caspase-3, -6, and -7), which in turn execute apoptosis by cleaving cellular proteins following specific aspartic acid residues. Activation of Fas and tumour necrosis factor (TNF) receptor 1 by Fas ligand (FasL) or TNF, respectively, leads to the activation of caspases-8 and -10. Cytochrome c released from damaged mitochondria is coupled to the activation of caspase-9. Mitochondria release multiple pro-apoptotic molecules, such as Smac/Diablo, apoptosis inducing factor (AIF), HtrA2 and EndoG, in addition to cytochrome c. Smac/Diablo binds to x-linked inhibitor of apoptosis protein (XIAP) which prevents it from inhibiting caspases. Taken from (Cellsignal 2008).

2.2 Induction of apoptosis by cytotoxic cells of the immune system

distinct ways to induce apoptosis in target cells (Barry and Bleackley 2002; Henkart 1985; Lieberman 2003; Russell and Ley 2002; Trapani and Smyth 2002). Via the receptor-mediated pathway involving FasL and via the granule-exocytosis pathway using cytotoxic granules. The two pathways can be distinguished by their requirement for calcium: the granule pathway is dependent on calcium, whereas the receptor pathway is not (Rouvier et al. 1993). In any case upon recognition of a target cell via a foreign peptide presented on a major histocompatibility complex (MHC) class I molecule, an immunological synapse is formed containing different domains (Bromley et al. 2001; Davis 2002; Huppa and Davis 2003; Stinchcombe et al. 2001; Stinchcombe and Griffiths 2003). In a secretory domain, mediators for cytotoxicity are released, whereas a signalling domain contains the T-cell receptor (TCR) and cluster of differentiation (CD)8 on the site of the CTL interacting with the MHC class I molecule of the target cell. The synapse is stabilised by adhesion molecules such as leukocyte function-associated antigen-1 (LFA-1) and on the site of the target cell intercellular cell adhesion molecule-1 (ICAM-1) (Goldstein et al. 2000). For the FasL receptor (also known as CD178) induction of cell death by engaging Fas (CD95) on the target cell the immunological synapse is required as well (Stinchcombe et al. 2001). Either the release of cytotoxic granules into the synapse or the interaction of FasL with Fas leads to apoptosis of the target cell.

Activation of receptors of the TNF family of death receptors on target cells by the respective ligands on CTLs initiates the classical apoptotic cascade in target cells by activation of caspases (reviewed in (Barry and Bleackley 2002)). Henkart et al. (1997) and Sarin et al. (1997) on the other hand were able to show that apoptosis induced via the granule-exocytosis pathway can also take place in the presence of caspase-inhibitors. In general, it is assumed that most of the killing of transformed or infected cells is achieved by releasing the contents of cytotoxic granules, whereas the receptor-mediated pathway is thought to be important for immune responses by regulating the elimination of self-reactive lymphocytes (Lieberman 2003; Van Parijs and Abbas 1996). However, this should not be understood as a complete separation of function of the two pathways.

Cytotoxic granules are specialised secretory lysosomes just found in cytotoxic cells (Griffiths and Isaaz 1993) containing a variety of effector molecules with different functions. The expression of these molecules is regulated in CTLs but constitutive in NK cells. Upon target cell recognition cytotoxic granules migrate to the cell surface along microtubules under the control of the microtubule-organising centre (MTOC), fuse with it and release their contents into the synapse (Kuhn and Poenie 2002). Cytotoxic granules in humans contain a family of serine proteases named granzymes, two membrane perturbing proteins perforin and granulysin (which is not present in rodents), the proteoglycan matrix protein serglycin, the perforin-inhibitor calreticulin, lysosomal enzymes such as cathepsins and also FasL (Bossi and Griffiths 1999).

Perforin is a key component of the granule-exocytosis pathway as the functions of granzymes

2 Introduction

are largely redundant but there is just perforin for delivering granzymes into target cells. There is a requirement for perforin to deliver granzymes into the target cells although the exact mechanism is still unknown. It is known so far that in high concentrations perforin can accumulate and polymerise in the target cell membrane in a calcium-dependent manner, where it can form pores leading to necrosis and osmotic cell death (Henkart 1985). The importance of perforin was demonstrated in perforin-deficient mice, which showed a significant immunodefficiency along with an impaired protection against tumours and viruses (Kagi et al. 1994; Lowin et al. 1994a, 1994b; Seki et al. 2002; Smyth et al. 2000). The common model that perforin forms pores in the plasma membrane of target cells through which granzymes can enter was challenged by the findings that granzymes are secreted in a way bound to the 250 kilo Dalton (kDa) serglycin molecule, which is too big to enter through rather small pores (Metkar et al. 2002). Furthermore, only sublytic concentrations of perforin, which do not form pores, are sufficient to allow granzymes to induce apoptosis (Froelich et al. 1996a). Granzyme (Gr) B can bind to cells and get internalised into endosomal compartments of target cells in the absence of perforin (Froelich et al. 1996b; Pinkoski et al. 1998; Shi et al. 1997). The execution of apoptosis depends on perforin in order to release GrB from endosomes into the cytosol (Froelich et al. 1996b). This model is supported by the fact that other endosomolytic agents as attenuated adenovirus (AdV) or bacterial toxins such as streptolysin O and listeriolysin O can substitute for this function of perforin (Browne et al. 1999; Froelich et al. 1996b). These agents are useful research tools as isolated perforin is very unstable.

Granzymes are highly specific cytotoxic proteases (Barry and Bleackley 2002; Lieberman 2003; Trapani and Smyth 2002) processed by cleaving by cathepsin C from inactive pro-enzymes to active enzymes either in the granules or on their way to the cell surface. In the acidic pH of the granules granzymes are inactive. Humans possess granzymes A, B, H, K, and M, whereas mice possess A, B, C, D, E, F, G, K, L, M, and N (Grossman et al. 2003).

GrB is so far the best investigated granzyme (Lord et al. 2003). It was proposed that the mannose-6-phosphate receptor (MPR300), which targets newly synthesised granzymes to cytotoxic granules, is also the receptor responsible for GrB uptake into target cells and for rejection of allogeneic cells (Motyka et al. 2000; Veugelers et al. 2006). However, an MPR300-independent pathway was also described (Trapani et al. 2003) and it was shown with mice deficient for MPR300 that CTL-mediated apoptosis could still take place. Thus, MPR300 is not essential for apoptosis and also not essential for rejection of allogeneic cells (Dressel et al. 2004a, 2004b). In 2005 it became evident that negatively charged cell surface heparan sulphate (HS) proteoglycans are involved in binding and uptake of GrB into target cells (Bird et al. 2005; Kurschus et al. 2005; Raja et al. 2005; Shi et al. 2005). After uptake into target cells GrB is able to trigger apoptosis fast and efficiently (Heusel et al. 1994) either in a caspase-dependent or in a caspase-independent way (Sarin et al. 1997; Trapani et al. 1998). It is a serine

2.2 Induction of apoptosis by cytotoxic cells of the immune system

protease cleaving after aspartic acid residues like caspases do, but caspase-3 cleaves about 10 times more cellular substrates than GrB (Lüthi and Martin 2007). Due to the similar preference for aspartic acid residues, it is not surprising, that GrB can directly activate caspases-3, -6, -7, -8, -9, and -10 (Darmon et al. 1995; Duan et al. 1996; Fernandes-Alnemri et al. 1996; Martin et al. 1996; Medema et al. 1997; Quan et al. 1996). Activation of pro-caspases-8 and -3 by GrB unleashes the classical apoptotic cascade via the extrinsic pathway (Atkinson et al. 1998; Medema et al. 1997; Metkar et al. 2003). GrB can activate the intrinsic apoptotic pathway by direct or indirect cleavage of BID (Alimonti et al. 2001; Barry et al. 2000; Darmon et al. 1995; Heibein et al. 2000; Sutton et al. 2000). Subsequently, pro-apoptotic factors are released from mitochondria and lead to apoptotic DNA fragmentation (Li et al. 2001). Remarkably, GrB can also induce apoptosis in the presence of capsase-inhibitors, whereby all known non-caspase substrates of GrB are downstream caspase substrates, so that GrB takes on the role of caspases in such a case (Andrade et al. 1998; Froelich et al. 1996a; Zhang et al. 2001). The nuclear apoptosis pathway can directly be activated by cleavage of ICAD, so that CAD becomes activated and can cause the prominent oligonucleosomal DNA damages mentioned above (Sharif-Askari et al. 2001; Thomas et al. 2000). GrB also targets the cytoskeleton, namely α-tubulin (Adrain et al. 2006; Goping et al. 2006). The function of GrB in comparison to perforin is redundant as mice genetically deficient for the GrB gene (*Gzm b*) are hardly less susceptible to tumours or viral infections than wild type (Wt) mice (Zajac et al. 2003).

GrA is a tryptase, which cleaves substrates after lysine or arginine residues and induces apoptosis in a caspase-independent way (Beresford et al. 1999; Fan et al. 2003; Shresta et al. 1999), which is slower than GrB-induced apoptosis (Masson et al. 1986). It still shows characteristics of apoptosis like condensation of chromatin, nuclear fragmentation by destruction of lamins in the nuclear envelope, loss of mitochondrial membrane potential $\Delta\Psi$ and the externalisation of phosphatidylserines (PSs), normally located in the inner layer of the cell membrane (Beresford et al. 1999; Zhang et al. 2001). The type of DNA damage caused by GrA is different from the one caused by GrB as it consists of single-stranded nicks, which cannot be labelled with terminal desoxynucleotidyltransferase (TdT) but with Klenow polymerase. Those nicks are evoked by GrA-activated DNase (GAAD) also known as NM23-H1, which is inhibited by the SET complex (Fan et al. 2003). GrA destroys three members the SET complex and thereby activates GAAD (Martinvalet et al. 2005). Furthermore, GrA is able to completely degrade linker histone H1 and to cut off the tails from core histones (Zhang et al. 2001). Mice genetically deficient for the GrA gene (*Gzm a*) are highly susceptible towards certain viral infections (Müllbacher et al. 1996). A combined loss of GrA and GrB leads to major defects in cellular cytotoxicity comparable to the one in perforin-deficient mice (Müllbacher et al. 1999; Pham et al. 1996).

GrC is encoded in a cluster downstream of the gene for GrB on chromosome 14, together

2 Introduction

with granzymes G, D, E, and F. In contrast to GrA and GrB, it is not so highly expressed after antigenic stimulation of CTLs but it evokes single-stranded nicks. The other similarity to GrA is that it induces apoptosis completely caspase-independent. Cell death is especially characterised by mitochondrial swelling and rapid externalisation of PS (Johnson et al. 2003). The apoptotic pathway triggered by GrC is clearly distinct from that unleashed by GrA and GrB.

It could be demonstrated by Andrade et al. (2007) that granzymes act in a synergistic fashion. Upon induction of apoptosis, e.g. in AdV-infected cells by cytotoxic cells, GrB can get inhibited by the adenoviral 100K assembly protein. Another granzyme, namely GrH, can then directly cleave the adenoviral 100K assembly protein to release GrB and induce apoptosis.

In cytotoxic granules basic granzymes are bound in a non-covalent fashion to the negatively charged proteoglycan serglycin. It consists of alternating serine and glycine residues on a small 17 kDa (Stevens et al. 1988) core protein making up a 250 kDa molecule, to which approximately 30–50 granzyme molecules can bind (Metkar et al. 2002; Raja et al. 2002). Serglycin-deficient mice are impaired in their ability to properly store GrB but not GrA or perforin in cytotoxic granules. CTLs derived from these mice were not impaired in induction of apoptosis as determined by [^{3}H]-Thymidine-release assays (Grujic et al. 2005). A role for serglycin in regulating the kinetics of contraction of $CD8^+$ T-cell populations after virus infection was proposed (Grujic et al. 2008). Nevertheless, it was shown that serglycin-complexed as well as isolated GrB is able to induce apoptosis, if administered with just sublytic doses of perforin (Metkar et al. 2002).

2.3 Role of *Sulf1* and *Sulf2* on heparan sulphates and in the uptake of granzyme B into target cells

The different existing models for the uptake of GrB into target cells were already described in section 2.2 on page 9. Recently, the model involving heparan sulphate (HS) proteoglycans (PGs) gained special interest. HS, a strongly anionic linear polysaccharide, is consistently found on membrane-bound proteoglycans, syndecans and glypicans. HS is ubiquitously expressed on most mammalian cells and belongs to the most abundant glycans. The presence of HS is essential to vertebrate-life as embryos deficient for HS are incapabable of life (Lin et al. 2000). A large variety of ligands can interact with HS including growth or morphogenic factors (Bernfield et al. 1999). GrB was described to interact with HS (Metkar et al. 2002). The importance of HS is further enlightened by its role in various signalling pathways, e.g. sonic hedgehog (Dierker et al. 2009), fibroblast growth factor (Lamanna et al. 2008), and Wnt signalling (Dhoot et al. 2001). Dhoot and colleagues discovered that the signalling function of cell surface HSPGs depends on their sulphation pattern and that this pattern is modulated by 6-O-endosulphatases like QSULF1. The mammalian, hSULF1 and hSULF2, and murine orthologues, mSULF1 and mSULF2, were described by Morimoto-Tomita et al. (2002) and Habuchi et al. (2000), respectively.

It has been demonstrated previously, that the enzyme specificity of sulphatase 1 and sulphatase 2 is not restricted to di- and tri-sulphated 6-S disaccharide units within the HS chain. Additionally, the *Sulf1* and *Sulf2* genes have an impact on the 6-O-sulphotransferase activity (Lamanna et al. 2008). The 6-O-sulphation of cell surface PGs shows an increase in monosulphated 6-S disaccharides in *Sulf1*-deficient mouse embryonic fibroblasts (MEFs) of about 10 % in comparison to Wt MEFs. The mono-sulphated 6-S disaccharides are increased by about 50 % in the *Sulf2* and about 30 % in the *Sulf* double-deficient MEFs in comparison to Wt cells. The increase for di- and tri-sulphated 6-S disaccharides is similar for both single *Sulf*-deficient cell lines; for the *Sulf* double-deficient cell line Lamanna et al. could show 60–80 % increase of sulphation (Lamanna et al. 2006, 2008).

The increased sulphation pattern of HS caused by the knock-out of *Sulf1* and *Sulf2* drastically increased the negative charge of cell surface HSPGs on MEFs. The exchange of the basic protein GrB complexed with anionic serglycin-PGs in the immunological synapse to anionic HSPGs on the target cell surface was described to be an electrostatic exchange (Raja et al. 2005). It might therefore be, that the deficiency of *Sulf1* and *Sulf2* increases the electrostatic exchange of GrB and with this binding to HS and maybe also the uptake into these MEFs. Consequently, *Sulf1* and *Sulf2* genes might contribute to the control of target cell susceptibility to GrB and CTL-induced apoptosis.

2.4 The role of heat shock protein 70 in apoptosis

2.4.1 Heat shock protein 70

A mild heat shock can induce the stress response system including the expression of heat shock proteins leading to a transient state of thermotolerance (Li and Werb 1982). Additionally, also physiological, chemical, and environmental stresses can induce a rapid expression of heat shock proteins (Ciocca et al. 1992; Santoro 2000; Trautinger et al. 1999). One of the most prominent heat shock protein families is the heat shock protein 70 family (Günther and Walter 1994).

A connection between the major heat shock protein of 70 kDa (HSP70) and the immune system was assumed first when it was discovered, that three HSP70 genes are encoded within the MHC on the short arm of chromosome 6 (Sargent et al. 1989; Wurst et al. 1989).

Various members of the highly conserved HSP70 protein family are known and localised in distinct intracellular compartments. All members of the HSP70 family are molecular chaperones

2 Introduction

but just a few, as the MHC-linked HSP70-1 (*HSPA1A*) and HSP70-2 (*HSPA1B*) genes, are strongly induced upon heat shock (Dressel and Günther 1999; Dressel et al. 1996). Other members such as the cytosolic heat shock cognate 70 (HSC70), the glucose regulated protein (GRP)78 located in the endoplasmic reticulum, and the mitochondrial HSP75 are constitutively expressed. They exert important physiological functions such as prohibiting premature folding or even misfolding of newly synthesised proteins, or accompanying their transport into specific intracellular compartments. This often involves unfolding and refolding of proteins to cross membranes. Furthermore, chaperones also facilitate the assembly of multi-protein complexes (Gething and Sambrook 1992; Hartl and Hayer-Hartl 2002; Mayer et al. 2001) and newer data imply that some proteins even need continuous help of chaperones for fulfilling their normal functions (Pratt and Toft 2003).

The human HSP70 consists of 641 amino acids (aa), whereby it possesses an adenosine triphosphate (ATP)ase domain in the N-terminal region and a peptide-binding domain towards the C-terminal region.

2.4.2 Heat shock protein 70 in apoptosis

The cytoprotective effect provided by heat shock proteins could be shown in many experimental systems (reviewed in: (Beere 2004, 2005; Jäättelä 1999; Samali and Cotter 1996; Sreedhar and Csermely 2004)). A protective effect in cells exposed to cytotoxic mechanisms of the immune system could be demonstrated already quite early. Initially it was shown that a heat shock confers resistance of tumour cells against TNF-α (Gromkowski et al. 1989; Jäättelä et al. 1989; Kusher et al. 1990; Sugawara et al. 1990), CTLs (Geginat et al. 1993; Sugawara et al. 1990), and monocytes (Jäättelä and Wissing 1993). Later it was demonstrated that also transfection with heat shock genes including the *HSP70* gene can render cells resistant towards TNF-α (Jäättelä et al. 1992, 1998) and other non-immunological stimuli such as UV-irradiation (Simon et al. 1995), ceramide (Ahn et al. 1999), ischemia (Hoehn et al. 2001), or serum withdrawal (Ravagnan et al. 2001).

In the beginning, stress resistance provided by heat shock proteins was mainly attributed to their function as chaperones, which prevent misfolding and aid in re-folding of denatured proteins after stress. Later, it has become obvious that heat shock proteins, including HSP70, interfere with several specific steps of different apoptotic pathways (Dressel and Demiroglu 2006). It was described that HSP70 can prevent the formation of a functional apoptosome by blocking the recruitment of caspase-9 (Beere et al. 2000; Saleh et al. 2000). Furthermore, HSP70 can suppress the c-Jun-N-terminal kinase (JNK) (stress kinase), which is part of a pro-apoptotic signalling pathway (Bienemann et al. 2008; Gabai et al. 2002, 1997; Mosser et al. 2000). HSP70 can prevent the release of cytochrome c from mitochondria (Bivik et al. 2007; Mosser et al. 2000) by inhibiting pro-apoptotic molecules such as the BCL-2 family member

2.4 The role of heat shock protein 70 in apoptosis

BAX and thereby stabilises the mitochondrial membrane (Stankiewicz et al. 2005). Moreover, it prevents the nuclear import of AIF (Chaitanya and Babu 2008; Gurbuxani et al. 2003) and inhibits apoptosis by stabilising lysosomes (Dudeja et al. 2009). These are some examples to indicate that HSP70 is able to interfere at several steps of the apoptotic cascade to abrogate cell death. In conclusion, in the stress response system intracellular HSP70 can provide cellular protection.

In accordance with these results HSP70 is found to be overexpressed in many human tumours (Mosser and Morimoto 2004). In some types of tumours, e.g. breast cancer, this correlates with a poor prognosis and resistance against therapy (Garrido et al. 2001; Jäättelä 1999; Jolly and Morimoto 2000). However, this does not seem to be the case for all kinds of tumours as in osteosarcomas and kidney cell sarcomas the overexpression of HSP70 was found to be associated with a rather good prognosis (Santarosa et al. 1997; Trieb et al. 1998). This might be due to immunological functions of HSP70.

While intracellular HSP70 in the stress response system mainly protects cells from apoptotic stimuli, extracellular HSP70 can act as an immunological danger signal to activate cells of the innate and adaptive immune system.

HSP70 and other heat shock proteins chaperone antigenic peptides in the cytosol and in the endoplasmic reticulum and prevent their degradation before they are loaded on MHC class I molecules (Srivastava et al. 1998, 1994). Therefore, preparations of HSP70 and other heat shock proteins from virus-infected or tumour cells contain the antigenic repertoire of these cells and can be used as vaccines to stimulate an antigen-specific immune response (Udono and Srivastava 1993). This is particularly efficient because heat shock protein-peptide complexes are bound to receptors, such as CD91 and lectin-like oxidised low-density lipoprotein receptor-1 (LOX-1), on professional antigen presenting cells (APCs) and are internalised (Basu et al. 2001; Binder et al. 2000; Castellino et al. 2000). The chaperoned peptides are then channeled into the MHC class I presentation pathway and elicit a strong CTL response. In addition, binding of HSP70 to other receptors including Toll-like receptor (TLR)2/4, CD14, CD36, and CD40 can induce the release of pro-inflammatory cytokines and initiate innate immune reactions (Srivastava 2002a, 2002b).

In addition, HSP70 can be translocated to the plasma membrane of stressed cells or is released together with parts of the intact membrane of these cells (exosomes) and stimulates macrophages (Vega et al. 2008). When HSP70 is expressed on the cell surface of tumour cells it can function as a recognition structure for NK cells (Gastpar et al. 2004; Multhoff et al. 1997, 1995). Furthermore, stimulation of NK cells with HSP70 or the HSP70-derived peptide TKD can enhance NK cell-mediated cytotoxicity towards HSP70 plasmamembrane-positive tumours (Multhoff 2002; Multhoff et al. 1997, 1999, 2001).

In summary as demonstrated in figure 2.2 on the next page, on the one side HSP70 protects cells by inhibiting apoptosis at many different steps and on the other side it can activate cells

2 Introduction

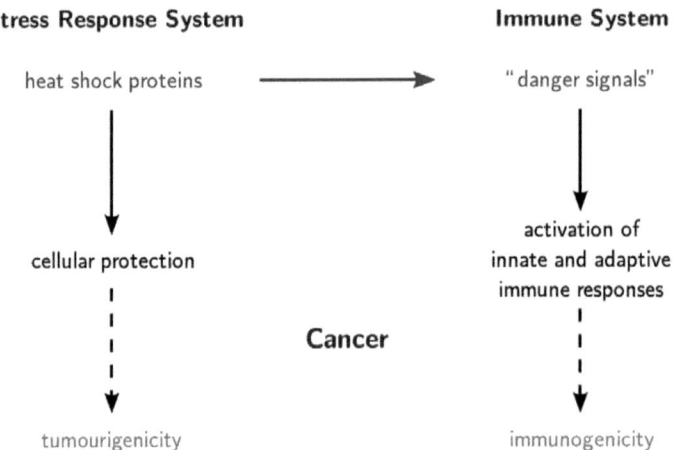

Figure 2.2: Role of HSP70 in the stress response and in the immune system HSP70 has diverse functions in the stress response and in the immune system. In the stress response system intracellular HSP70 acts as chaperone and protects the cell from cytotoxic stimuli. In tumour cells this protective function of intracellular HSP70 can promote tumourigenicity. On the other hand, in the immune system extracellular HSP70, released from apoptotic or necrotic cells or present on the cell surface of some tumours, acts as "danger signal" and can activate cells of the innate and adaptive immune response. Thus, HSP70 can incerase immunogenicity. In summary, depending on its location, intracellularly or extracellularly, HSP70 has two opposing functions, which could promote or diminish cancer, respectively.

of the innate and adaptive immune response when it is released from dying cells or is used as a vaccine.

We were therefore interested in the role of HSP70 in apoptosis induced by cytotoxic cells of the immune system via the granule-exocytosis pathway, which is distinct different from other apoptotic pathways. Would HSP70 be protective against cytotoxic cells although it is able to activate them?

We observed that the rat myeloma cell line 210-RCY3-Ag1.2.3 (Y3) did not express HSP70, even not after heat shock, but that heat shock rendered these cells resistant towards CTLs. The transfer of recombinant HSP70 into Y3 cells abolished the resistance against CTLs (Dressel et al. 2000). This result suggested that HSP70 can improve cell death induced by CTLs in heat-shocked target cells.

The effects of HSP70 were further analysed in the human melanoma cell line Ge. For conditional overexpression of HSP70, the cell line was transfected with a rat *Hsp70* gene under the control of a tetracycline-inducible promoter (Ge-tet cells) (Dressel et al. 1999). For permanent overexpression of HSP70, Ge cells were retrovirally transduced with the same rat *Hsp70* gene

(Ge-Hsp70 cells) (Dressel et al. 2003).

The acute overexpression of HSP70 for 24 hours (hrs) in Ge-tet cells increased the lysis of tumour cells by CTLs using the granule-exocytosis pathway. This effect was neither caused by an increase in MHC class I expression on the cell surface of target cells nor by a function of HSP70 in antigen processing (Dressel et al. 1999). This is in contrast to the findings of Wells et al. (1998) describing that a transfection of the mouse melanoma cell line B16 with HSP70 increased the MHC class I expression and improved thereby recognition and lysis by CTLs. Thus, we assumed that in cytotoxic granule-mediated apoptosis acutely overexpressed HSP70 improves the function of proteins involved in this process. Interestingly, the constitutive overexpression of HSP70 in Ge-Hsp70 cells did not improve susceptibility towards CTLs (Dressel et al. 2003), which implies that the increase in susceptibility does not depend on the level of HSP70 expression but rather on the availability of additional HSP70. Long-term in contrast to short-term overexpression of HSP70 in Ge-tet cells did also not increase the lysis by CTLs (Dressel et al. 2003). This can be explained by a compensation within the chaperone network down-regulating the constitutively expressed HSC70 upon prolonged overexpression of HSP70 (Dressel et al. 2003). Thus, the increase of susceptibility to CTLs seen in cells acutely overexpressing HSP70 seems to be mediated by HSP70 proteins that are not occupied in physiological functions and are free to chaperone molecules involved in the execution of apoptosis. This concept is supported by the findings that the chaperone HSP60 binds to pro-caspase-3 and is required for its activation (Samali et al. 1999; Xanthoudakis et al. 1999). Thus, HSP60 acts as a pro-apoptotic protein in this context and HSP70 can also carry-out pro-apoptotic functions under certain circumstances. Furthermore, HSP70 can stabilise the function of CAD, which is able to induce DNA fragmentation in the nucleus, although these findings were reported from TCR-induced T-cells, they show that HSP70 is able to chaperone pro-apoptotic proteins (Liu et al. 2003).

However, so far, nothing is known about the molecular mechanisms that confer the increased sensitivity of Ge-tet cells to CTLs after acute overexpression of HSP70. Interestingly, it was described that HSP70 on the cell surface of target cells can directly interact with GrB and also mediate its uptake (Gross et al. 2003b). This suggests that HSP70 might improve specifically the GrB-induced apoptosis.

2.5 Aims

It was the aim of this project to further analyse the role of HSP70 and sulphatase 1 and 2 genes in apoptosis induced by cytotoxic cells of the immune system via the granule-exocytosis pathway.

Previous results had indicated that acute overexpression of HSP70 can augment the lysis of

2 Introduction

Ge melanoma cells by CTLs using the granule-exocytosis pathway. The molecular mechanism behind this phenomenon is unknown. To further elucidate these pro-apoptotic effects two approaches were followed: (1) To determine whether the cell-death promoting effect of acutely overexpressed HSP70 results from a specific regulation of genes, e.g. up-regulation of genes encoding for pro or down-regulation of genes encoding for anti-apoptotic proteins, an expression profiling experiment should be performed in the Ge-tet system. (2) To reduce the complexity of killing in the granule-exocytosis pathway, effects of specific effector molecules should be analysed. Priority should be given to GrB because this effector protease is known to interact with HSP70. It should be analysed which key steps in GrB-mediated apoptosis might be affected by HSP70 including the activation of caspases, the loss of the mitochondrial membrane potential $\Delta\Psi$, and DNA fragmentation.

In the second part of the project the role of two other genes in CTL-mediated apoptosis should be investigated. It is known from our work that HSPGs are involved in the uptake of cytotoxic effector molecules such as GrB and control the efficiency of CTL and GrB-induced apoptosis (Raja et al. 2005). Therefore, we planned to investigate the effect of increased sulphation of HSPGs on CTL and GrB-induced apoptosis using cells from mice deficient for the *Sulf1* and *Sulf2* genes.

Together the results of this project might contribute to the understanding of factors that control the susceptibility of tumour cells to cytotoxicity mediated by CTLs in the granule-exocytosis pathway.

3 Materials

3.1 Antibodies and dyes

Primary antibodies are listed in table 3.1, secondary antibodies and isotype controls in table 3.2 on the next page and dyes in table 3.3 on the following page. Antibodies were used for flow cytometry and for western blot analysis. Dyes were used for flow cytometric analyses.

Table 3.1: Antibodies Antibodies are listed with specificity, fluorescence labelling, the clone name, the isotype, and also the provider. Antibodies were either conjugated with fluoresceinisothiocyanate (FITC), red phycoerythrin (RPE), tri-colour (TC), horseradish peroxidase (HRP), or biotin or not conjugated at all. Some antibodies are directed against cell surface structures named CD.

Antibody	Clone; Isotype	Provider
anti-human CD3 FITC-conjugated	clone MEM57; isotype: mouse IgG_{2a}	ImmunoTools
anti-human CD4 RPE-conjugated	clone S3.5; isotype: mouse IgG_{2a}	Caltag Laboratories
anti-human CD8 TC-conjugated	clone 3B5; isotype: mouse IgG_1	Caltag Laboratories
anti-human CD16 TC-conjugated	clone 3G3; isotype: mouse IgG_{2a}	Caltag Laboratories
anti-human CD56 RPE-conjugated	clone MEM-188; isotype: mouse IgG_{2a}	ImmunoTools
anti-human CD94 FITC-conjugated	clone HP-3D9; isotype: mouse IgG_1	BD Pharmingen
anti-mouse $H2K^b$ RPE-conjugated	clone CTK^b; isotype: mouse IgG_{2a}	Caltag Laboratories
anti-HSP70	clone C92F3A-5; isotype: mouse IgG_1	Stressgen
anti-HSC70	clone 1B5; isotype: rat IgG_{2a}	Stressgen

Continued on next page

3 Materials

Table 3.1 – continued from previous page

Antibody specificity	Clone; Isotype	Provider
anti-heparan sulphate	clone F58-10E4; isotype: mouse IgM	Seikagaku Corporation
anti-heparan sulphate stub	clone 3G10; isotype: mouse IgG_{2b}	Seikagaku Corporation
anti-CAR	clone E(mh)1; isotype: mouse IgG_1	Santa Cruz Biotechnology
anti-Integrin α_v (CD51) RPE-conjugated	clone RMV-7; isotype: rat IgG_1	BD Pharmingen
anti-active caspase-3 FITC-conjugated	clone C92-605; isotype: rabbit IgG	BD Pharmingen
anti-caspase-8	clone 1C12; isotype: mouse IgG_1	Cell Signaling
anti-cytochrome c	clone 7H8.2C12; isotype: mouse IgG_{2b}	BD Pharmingen

Table 3.2: Secondary antibodies and isotype controls Listed are secondary antibodies with name and flurorescence labelling and provider.

Name	Provider
goat anti-mouse IgM FITC-conjugated	Dianova
rabbit anti-goat IgG (H+L) HRP-conjugated	Dianova
goat anti-mouse IgG HRP-conjugated	Dianova
goat anti-rat IgG HRP-conjugated	Dianova
mouse IgG_{2a} RPE-conjugated	Caltag Laboratories

Table 3.3: Dyes Dyes are listed with specificity, fluorescence labelling and also the provider.

Dye name	Specificity	Provider
Annexin V-FITC	to stain phosphatidylserine residues on the cell surface	BD Pharmingen
JC-1	to determine change in mitochondrial membrane potential	Invitrogen
DiD	to stain whole cells; from Vybrant Multicolor cell-labelling kit	Invitrogen

3.2 Primers

All primers for quantitative real-time PCR (qRT-PCR) were designed with Primer3 (Rozen and Skaletsky 2000) and ordered at Invitrogen. They were designed, that the product size was between 70 and 90 base pairs (bp), the primer length about 20 bases and the melting temperature at about 65 °Celsius (C). The primer sequences were preferentially located in the 3' untranslated region (UTR) of the respective gene in order to include the oligo sequence used for the microarray analysis. Specificity was checked by blasting primer sequences against human transcripts at (NCBI 2008). The primers for human *GAPDH* were purchased from Qiagen. Both, forward and reverse primer, detect transcript NM_002046 of human *GAPDH* with an amplicon length of 119 bp.

Table 3.4: Primers for qRT-PCR Given are the gene names in capital letters, the addition, whether it is the forward (-For) or reverse (-Rev) primer sequence, the sequence itself and the reference sequence (Refseq). All sequences are human sequences for the respective genes, except for HSP70, which is rat. All primer sequences are depicted in 5' to 3' direction.

Name	Sequence	Refseq
ASNS-For	cgtgttggatggggactgtg	NM_001673
ASNS-Rev	ttttcacacccaagttagcctga	
BCL10-For	gtgtgccaccatgcctcact	NM_003921
BCL10-Rev	aagaccagcctggccaacat	
CASP8-For	ccccaaacttgctttatgccttc	NM_001080124
CASP8-Rev	ccccagagcattgttagcaaaa	
EIF5-For	ggacctgacagagcccatgc	NM_001969
EIF5-Rev	tcaggcaaggagttcatgagga	
GRB10-For	cgttttcagggaatgcagaagg	NM_001001549
GRB10-Rev	cagaatgaagcaaagcacatgga	
Hsp70-For	cgaggaggtggattagaggcttt	NW_047597.1
Hsp70-Rev	gtgcaccagcagccatcaag	
JUND-For	gtctcggctgcccctttgta	NM_005354
JUND-Rev	aaaggaaaggcagggtttgagg	
SAPS3-For	tgccttttaacccattcaccaaa	NM_018312
SAPS3-Rev	cagattgctctgaaatgttcattgg	
STC2-For	gaagtcagggcggctggatt	NM_003714
STC2-Rev	tgcctcctctccacccttctc	
TBL1XR1-For	gcactattgtgaaaaggagcaacg	NM_024665
TBL1XR1-Rev	cacagaatgatggacacttcgaga	
THOC4-For	aggacccaggcgtctcctct	XM_001134346

Continued on next page

3 Materials

Table 3.4 – continued from previous page

Name	Sequence	Refseq
THOC4-Rev	atccatcattggccgcacag	
TM4SF1-For	gcaaacgatgtgcgatgctt	NM_014220
TM4SF1-Rev	agggctgccacaatgacaca	
TXNRD1-For	gtttccgtgcccaaatccaa	NM_003330
TXNRD1-Rev	aagcacaggacacgcaggtg	

3.3 Chemicals and reagents

All chemicals and other reagents for cell culture and experiments are listed in table 3.5 together with the name of the provider.

Table 3.5: Chemicals, antibiotics, peptides, proteins, enzymes, reagents used in cell culture, radioactive substances, and other reagents.

Chemical/Reagent	Provider
[^{3}H]-Thymidine	Amersham Biosciences
100 bp DNA ladder	Genecraft
β-mercaptoethanol	Sigma-Aldrich Chemie GmbH
Acetic acid	Merck Biosciences GmbH
Acetonitrile	VWR International GmbH
Acrylamid 40	Roth GmbH & Co.
Agarose	Invitrogen
Ampicillin sodium salt	Roth GmbH & Co.
Ammonium chloride (NH$_4$Cl)	Merck Biosciences GmbH
Ammonium persulphate (APS)	Serva Electrophoresis GmbH
Anti-biotin microbeads	Miltenyi Biotec
Biocoll	Biochrom AG
Borate	Merck Biosciences GmbH
Bovine serum albumine (BSA), Fraction V	Merck Biosciences GmbH
Bromphenol blue	Merck Biosciences GmbH
Calcium chloride (CaCl$_2$)	Merck Biosciences GmbH
Chloroform	Merck Biosciences GmbH
Concanavalin A (Con A)	Sigma-Aldrich Chemie GmbH
Diaminobenzoide (DAB)	Roth GmbH & Co.
Dimethylsulfoxide (DMSO)	Merck Biosciences GmbH

Continued on next page

3.3 Chemicals and reagents

Table 3.5 – continued from previous page

Chemical/Reagent	Provider
dNTPs	Genecraft
Doxycycline Hyclate	Sigma-Aldrich Chemie GmbH
Dulbecco's modified Eagle's medium (DMEM)	Biochrom AG
Dithiotreitol (DTT)	Roth GmbH & Co.
Ethanol (EtOH) 99%	GeReSo GmbH
Ethidium bromide	Merck Biosciences GmbH
Ethylendiamine tetraacetic acid (EDTA)	Roth GmbH & Co.
Ethylene glycol tetraacetic acid (EGTA)	Roth GmbH & Co.
FACSflow	Becton Dickinson GmbH
Fetcal calf serum (FCS)	Biochrom AG
Glycine	Roth GmbH & Co.
GrB-Alexa 488	provided by Prof. Froelich
Heparinases II and III	provided by Prof. Dierks
High molecular weight standard mixture for SDS Gel electrophoresis (SDS-6H)	Sigma-Aldrich Chemie GmbH
Human GrB	Axxora
4-(2-hydroxyethyl)-1-piperazineethanesulfonic acid (Hepes)	Sigma-Aldrich Chemie GmbH
Hydrochloric acid (HCl)	Merck Biosciences GmbH
Isoamylalcohol	Merck Biosciences GmbH
Isopropanol	Merck Biosciences GmbH
Luminol	Biomol/Stressgen
Magnesium chloride ($MgCl_2$)	Merck Biosciences GmbH
Methanol	Merck Biosciences GmbH
MMLV-reverse transcriptase	Promega GmbH
$Na_2\ ^{51}CrO_4$	Hartmann Analytic GmbH
N-Lauroylsarcosine 20%	Sigma-Aldrich Chemie GmbH
Optiphase Supermix	PerkinElmer LAS GmbH
Para-hydroxycoumarine acid	Sigma-Aldrich Chemie GmbH
Paraformaldehyde (PFA)	Merck Biosciences GmbH
Polyethylene glycol (PEG)	Serva Electrophoresis GmbH
Penicillin	Sigma-Aldrich Chemie GmbH
Perhydrol 30 % H_2O_2	Merck Biosciences GmbH
Phenol	Biomol/Stressgen

Continued on next page

3 Materials

Table 3.5 – continued from previous page

Chemical/Reagent	Provider
Phosphate-buffered saline (PBS)	Biochrom AG
Phosphoric acid	Merck Biosciences GmbH
Ponceau S	Sigma-Aldrich Chemie GmbH
Potassium bicarbonate ($KHCO_3$)	Merck Biosciences GmbH
Propidium iodide (PI)	Sigma-Aldrich Chemie GmbH
Proteinase K	Merck Biosciences GmbH
Pyrophosphatase, inorganic	New England Biolabs GmbH
Pyruvic acid	Sigma-Aldrich Chemie GmbH
Random primers	Promega GmbH
Recombinant mouse interleukin-2 (IL-2)	R & D Systems GmbH
RNase A	Roche Diagnostics GmbH
RNase-out	Promega GmbH
RNasin plus RNase inhibitor	Promega GmbH
Saponin	Roth GmbH & Co.
SIINFEKL (Ovalbumin amino acid (aa) 257–264)	Bachem AG
Sodium dodecyl sulphate (SDS)	Roth GmbH & Co.
Sodium bicarbonate ($NaHCO_3$)	Sigma-Aldrich Chemie GmbH
Sodium chloride (NaCl)	Roth GmbH & Co.
Sodium hydroxide (NaOH)	Merck KGaA
Staurosporine	Roth GmbH & Co.
Streptomycin sulphate (10 mg/ml)	Sigma-Aldrich Chemie GmbH
SYBR green mix	Applied Biosystems
Taq-polymerase	Genecraft
N,N,N,N-tetramethyl-ethane-1,2-diamine (TEMED)	AppliChem GmbH
Tris(hydroxymethyl)-aminomethane (Tris)	Roth GmbH & Co.
Triton X-100	Sigma-Aldrich Chemie GmbH
Trizol	Invitrogen
Trypan blue	Sigma-Aldrich Chemie GmbH
Trypsin	Biochrom AG
Tween-20	Sigma-Aldrich Chemie GmbH
Xylene cyanol	Merck Biosciences GmbH

3.4 Kits

All kits used for micorarray or other experiments are listed in table 3.6.

Table 3.6: Kits

Kit name	Provider
RNA 6000 Pico Chip Kit	Agilent Technologies
Low RNA Input Linear Amplification Kit, Plus, two color	Agilent Technologies
RNA Spike-In Kit, two color	Agilent Technologies
Gene Expression Hybridization Kit	Agilent Technologies
RNeasy Mini Kit	Qiagen
NK Cell Isolation Kit II (human)	Miltenyi Biotec

3.5 Buffers and stock solutions

All buffers and solutions were made with deionised water (filtered water with a Sartorius arium 611 UF) and either filtered sterile or autoclaved for 30 minutes (min) at 121 °C. Buffers and solutions used for the experiments are listed in table 3.7, whereby the commercially available buffers and solutions for the microarray are listed separately in table 3.8 on page 30.

Table 3.7: Buffers and stock solutions Listed are buffers, stock solutions. Components and the applications are given in brackets.

Name	Components
EDTA/PBS (for detaching cells):	1 mM EDTA
	in PBS
EDTA/Trypsin (for detaching cells):	0.05 % (w/v) Trypsin
	0.02 % (w/v) EDTA
	in PBS
EGTA/MgCl$_2$ (inhibiting granule exocytosis):	4 mM EGTA
	8 mM MgCl$_2$
	10 % (v/v) FCS
	in Hepes-buffered DMEM adjusted to pH 7.2
Erythrocyte lysis buffer:	155 mM NH$_4$Cl
	10 mM KHCO$_3$
	0.1 mM EDTA
	in water (H$_2$O)

Continued on next page

3 Materials

Table 3.7 – continued from previous page

Name	Components
MACS buffer (used with MACS columns):	0.5 % (w/v) BSA 2 mM EDTA in PBS
PFA (for fixation of cells):	1 % (w/v) PFA in PBS NaOH until solution gets clear adjusted to pH 7.2
PBS/PI/RNase A (for analysis of cell cycle and apoptosis):	10 µg/ml PI 100 µg/ml RNase A in PBS
Saponin/PBS (for permeabilising cells):	0.25 % (w/v) Saponin in PBS
Trypan blue (staining dead cells for counting):	0.2 % (w/v) Trypan blue in H_2O 4.25 % (w/v) NaCl in H_2O mix Trypan blue with 1 NaCl in a 4:1 ratio
Annexin-V binding buffer:	10 mM Hepes 140 mM NaCl 2.5 mM $CaCl_2$ in H_2O
Triton lysis buffer (for lysing cells in [51]Chromium-release assays):	10 % (v/v) Triton X-100 in PBS
TE lysis buffer (for lysing cells in [^{3}H]-Thymidine-release assays):	100 mM Tris 50 mM EDTA 1 % (v/v) Triton X-100 in PBS; adjusted to pH 8.0
SDS lysis buffer (for lysing cells and nuclei in [^{3}H]-Thymidine-release assays):	2 % (w/v) SDS 0.1 M NaOH in PBS
10 × running buffer (for SDS-polyacrylamide gel electrophoresis (PAGE)):	250 mM Tris 1.92 M Glycine in H_2O
SDS running buffer (for SDS-PAGE):	25 mM Tris 192 mM Glycine

Continued on next page

3.5 Buffers and stock solutions

Table 3.7 – continued from previous page

Name	Components
	0.1 % (w/v) SDS (heat inactivated for 1 hour at 65 °C)
	in H_2O
separating gel buffer (for SDS-PAGE):	1.5 M Tris/HCl pH 8.8
	0.4 % (w/v) SDS
	in H_2O
stacking gel buffer (for SDS-PAGE):	0.5 M Tris/HCl pH 6.8
	0.4 % (w/v) SDS
	in H_2O
sample buffer (reducing; for SDS-PAGE):	0.02 M Tris/HCl pH 8.0
	20 % (w/v) Glycerine
	2 % (w/v) SDS
	2 mM EDTA
	10 % (v/v) β-Mercaptoethanol
	0.1 % (w/v) Bromophenol blue
	in H_2O
separating gel (for SDS-PAGE):	5 ml Acrylamide 40 %
	5 ml separating gel buffer
	9.8 ml H_2O
	200 μl APS 10 %
	20 μl TEMED
stacking gel (for SDS-PAGE):	1 ml Acrylamide 40 %
	2.5 ml stacking gel buffer
	6.4 ml H_2O
	100 μl APS 10 %
	10 μl TEMED
Blotting buffer:	100 ml 10 x running buffer
	200 ml Methanol
	Ad 1 litre H_2O
PBS/Tween (washing and incubating of immunoblots):	0.05 or 0.1 % (v/v) Tween 20
	in PBS
DAB solution (for developing immunoblots):	25 mg DAB
	50 μl H_2O_2
	in 50 ml PBS/Tween

Continued on next page

3 Materials

Table 3.7 – continued from previous page

Name	Components
enhanced chemoluminescence (ECL) solution:	
4 ml of Solution A:	200 ml 0.1 M Tris/HCl pH 8.6
	50 mg Luminol
400 µl of Solution B:	11 mg para-Hydroxycoumarine acid
	dissolved in 10 ml DMSO
1.2 µl of 30 % H_2O_2	
TBE-buffer (for agarose gel electrophoresis):	12.11 g Tris
	5.14 g Borate
	0.37 g EDTA
	Ad 1 litre H_2O; adjusted to pH 8.3
TPE-buffer (for agarose gel electrophoresis):	80 mM Tris
	2 mM EDTA
	adjusted to pH 7.8 with phosphoric acid
10 × DNA loading dye (for agarose gel electrophoresis):	0.025 g Xylene cyanol
	0.025 g Bromophenol blue
	1.25 ml 10 % SDS
	12.5 ml Glycerol
	dissolved in 6.25 ml H_2O
Lysis buffer (for DNA fragmentation):	10 mM EDTA
	50 mM Tris pH 8.0
	0.5 % (w/v) Lauroylsarcosine
	0.5 mg/ml Proteinase K
	in H_2O

Table 3.8: Buffers and solutions for microarray Abbreviations: sodium chloride, sodium hydrogen phosphate, and EDTA (SSPE)

Name	Provider
2 × Hi-RPM Hybridisation buffer	Agilent Technologies
2 × Hybridisation buffer	Agilent Technologies
25 × Fragmentation buffer	Agilent Technologies
10 × Blocking Agent	Agilent Technologies
20 × SSPE buffer	Sigma-Aldrich Chemie GmbH

3.6 Cell lines, viruses and media

Mainly two types of cell lines were used: Ge, a human melanoma cell line, and MEFs.

Ge-tet cells contain a Tet-On system, in which the rat *Hsp70-1* gene is inducible by the addition of doxycycline. Ge-tra control cells just contain the transactivator domain of the Tet-On system, and therefore the addition of doxycycline does not induce *Hsp70-1*.

Sulf MEFs were kindly provided by Prof. Dierks (Bielefeld). *Sulf* MEFs from *Sulf*-deficient mice were immortalised by stable transfection with the pMSSVLT vector containing an SV40 fragment including a large T gene. The *Sulf* Wt clone was subcloned to generate two Wt clones with a higher H2K^b expression than the parental *Sulf* Wt provided by Prof. Dierks. The second set of independent *Sulf* MEFs indicated with an asterisk (*) were also provided by Prof. Dierks (Lamanna et al. 2008).

The generation of cytotoxic cells is described in section 4.2.1 on page 43.

Table 3.9: Cell lines Cell lines with species of origin and description are listed.

name	species	description
Ge	human melanoma	parental cell line
Ge-tra	human melanoma	transfected with transactivator domain only
Ge-tet-1 and 2	human melanoma	transfected with the rat *Hsp70-1* gene under the control of a tetracycline-inducible promotor (Dressel et al. 1999)
Ge-Hsp70-D	human melanoma	retrovirally transduced to permanently overexpress the rat *Hsp70-1* gene (Dressel et al. 2003)
Ge-TCR-D	human melanoma	retrovirally transduced to permanently overexpress β-chain of rat TCR
K-562	human chronic myelogeneous leukemia cell	NK cell target cell line; hardly any MHC class I molecules on the cell surface
Wt	mouse embryonic fibroblast	Wt control
Wt E9	mouse embryonic fibroblast	Wt subclone with higher H2K^b expression than the parental Wt clone
Wt F7	mouse embryonic fibroblast	Wt subclone with higher H2K^b expression than the parental Wt clone

Continued on next page

3 Materials

Table 3.9 – continued from previous page

name	species	description
Sulf1$^{-/-}$	mouse embryonic fibroblast	MEF derived from Sulf1-deficient mice (Lamanna et al. 2006)
Sulf2$^{-/-}$	mouse embryonic fibroblast	MEF derived from Sulf2-deficient mice
Sulf Dko	mouse embryonic fibroblast	MEF derived from Sulf1 Sulf2-deficient mice
Wt*	mouse embryonic fibroblast	Wildtype; Wt control for 2nd set of MEFs
Sulf1$^{-/-}$ *	mouse embryonic fibroblast	MEF derived from Sulf1-deficient mice; 2nd set of MEFs (Lamanna et al. 2008)
Sulf2$^{-/-}$ *	mouse embryonic fibroblast	MEF derived from Sulf2-deficient mice; 2nd set of MEFs
Sulf Dko*	mouse embryonic fibroblast	MEF derived from Sulf1 Sulf2-deficient mice; 2nd set of MEFs
RMA	mouse T-cell lymphoma	target cell line for mouse CTLs in cytotoxic assays
CHO	chinese hamster ovary cell	cell line that does not possess coxsackie and adenovirus receptor (CAR) on the cell surface
CHO A745	chinese hamster ovary cell	CHO mutant cell line that also does not possess HS on the cell surface
HeLa	human cervix carcinoma	cell line that served as positive control for CAR on the cell surface
Y3	rat myeloma	from LOU rat; does not express HSP70

QBI-AdV-green fluorescence protein (GFP) and QBI-AdV-β-galactosidase (AdV-β-gal) were kindly provided by Dr. Tim Seidler (University Hospital Göttingen, Molecular Cardiology). The recombinant AdV belongs to serotype 5 and was attenuated due to deletions of the E1 and E3 genes. AdV-GFP contains a GFP gene under the control of a eukaryotic cytomegalovirus (CMV) promoter. AdV-β-gal contains a Lacz gene also under the control of a eukaryotic CMV promoter. Both types of AdV were amplified in HEK293 cells, purified with cesium chloride ultracentrifugation and as last step dialysed against a sucrose buffer (Becker et al. 1994; Nyberg-Hoffman and Aguilar-Cordova 1999). In order to minimise contamination with Wt AdV a plaque purification was done and tested in a polymerase chain reaction (PCR) with primers specific for the deleted E1 region. The AdV-β-gal was used in some experiments, in which GrB-induced apoptosis was analysed. The biological activity of the AdV-GFP was 9.5

$\times 10^{10}$ plaque forming unit (pfu) per ml supernatant and of the AdV-β-gal was 9.6×10^{10} pfu per ml supernatant.

The different media used for cell culture are listed in table 3.10. They were sterile filtered using bottle top filters and a water suction pump.

Table 3.10: Cell culture media Listed are media with applications in brackets and components. For culture of cytotoxic effector cells, β-mercatptoethanol was additionally added to the DMEM.

Name	Components
DMEM (for cultivation in 10 % CO_2):	8.26 g DMEM per litre
	3.7 g $NaHCO_3$
	0.11 g Pyruvic acid
	100 000 U Penicillin
	100 mg Streptomycin sulphate
	10 % (v/v) FCS
	ad 1 litre H_2O
Hepes-buffered DMEM (for washing of cells):	8.26 g DMEM per litre
	4.77 g Hepes
	ad 1 litre H_2O; adjusted to pH 7.2
Hepes-buffered DMEM with FCS (for incubation without CO_2):	10 % (v/v) FCS in Hepes-buffered DMEM
DMEM 1 % BSA:	1 % (w/v) BSA in DMEM
freezing medium:	80 % (v/v) FCS
	20 % (v/v) DMSO
	mix 1:1 with Hepes-buffered DMEM immediately before freezing

3.7 Laboratory animals

Laboratory animals were bred in the animal facility of the faculty of human medicine of the University of Göttingen and are listed in table 3.11.

Table 3.11: Laboratory animals Strains with species and description are listed.

name	species	description
OT-I	mouse	transgenic for TCR (Hogquist et al. 1994)
C57Bl/6	mouse	MHC haplotype: $H2^b$

Continued on next page

3 Materials

Table 3.11 – continued from previous page

name	species	description
BN	rat	MHC haplotype: RT1^n
LOU	rat	MHC haplotype: RT1^u
BUF	rat	MHC haplotype: RT1^b

Spleens of OT-I mice were used in ^{51}Chromium and [^{3}H]-Thymidine-release assays. Spleens of rats from different strains (LOU, BUF, BN) were also dissected for the production of Con A supernatants.

3.8 Used laboratory equipment

All laboratory equipment, which was used, and the providers are listed in table 3.12. All equipment for use in cell culture was autoclaved for either 15 min at 134 °C or for 30 min 121 °C.

Table 3.12: Used laboratory equipment

Description	Name	Provider
autoclave:	High pressure steam sterilisator type A 40	Webeco
balances:	Vicon	Acculab
	BP 61	Sartorius
blotting chamber:		Biotec-Fischer
camera:	Polaroid CU-5	Polaroid GmbH
centrifuges:	Multifuge 1 L	Heraeus
	Multifuge 3 S-R	Heraeus
	3K30	Sigma
	Labofuge GL	Heraeus
	Minifuge GL	Heraeus
	5810R	Eppendorf
drying cupboard:	5000	Heraeus
electrophoresis chambers:	Easy-cast B1A	Peqlab
	H4	Bethesda Research Laboratories GmbH
	Prot-Resolv Maxi LC	Phase GmbH
FACS:	FACScan	Becton Dickinson
	FACSCalibur	Becton Dickinson

Continued on next page

3.8 Used laboratory equipment

Table 3.12 – continued from previous page

Description	Name	Provider
glass pipettes:	5 ml and 10 ml	Brand GmbH + Co KG
Hamilton syringe:	Hamilton microliter syringe	Hamilton Bonaduz AG
homogeniser:	Tenbroeck	Schütt Labortechnik GmbH
incubators:	Hera Cell 150	Heraeus
	B 5060 EK	Heraeus
magnetic stirrer/heater:	RH basic 2	IKA
for microarray:	Agilent technology platform	Agilent Technologies
	NanoDrop ND-1000 (RNA concentration)	Thermo Scientific
	2100 BioAnalyzer (RNA quality)	Agilent Technologies
	Microarray Scanner G2567AA	Agilent Technologies
for real-time PCR:	ABI 7500 Real-Time-PCR System	Applied Biosystems
microscopes:	No. 471202-9901	Zeiss
	No. 491220	Ernst Leitz GmbH
PCR-cycler:	Tpersonal	Biometra
pH meter:	CG-837	Schott
pipettes:	2.5 μl, 10 μl, 20 μl, 100 μl, 200 μl, 1000 μl, 300 μl 8-channel, Stepper	Eppendorf Research
	Automatic Sarpette	Desaga (Sarstedt-Group)
power packs:	2197 power supply	LKB Instruments
	Power Pack P25	Biometra
sonicator:	Sonopuls UW 2070	Bandelin electronic
sterile work bench:	Lamin Air HLB 2448	Heraeus
	type: CA/RE 5	Clean Air Deutschland GmbH
liquid scintillation counter:	1450 MicroBeta Trilux	Wallac
UV-light table:	Fluo link	Biometra
vortexer:	MS1 Minishaker	IKA
water bath:	Julabo U3	Omnilab-Krannich
	Thermocycler 60	Biomed
water treatment plant:	arium 611 UF	Sartorius AG
Western blot detection:	Intas Chemilux Entry	Intas Science Imaging Instruments

3.9 Disposable plastic ware and other disposables

All disposable plastic ware, which was used, and their providers are listed in table 3.13.

Table 3.13: Disposable plastic ware and other disposables

Name	Provider
1.5 ml tubes	Sarstedt
13 ml tubes	Sarstedt
15 ml tubes	Sarstedt
50 ml tubes	Sarstedt
96-well Wallac plates	PerkinElmer
top seal for Wallac plates	PerkinElmer
96-well plates for real-time PCR	Applied Biosystems
top seal for real-time PCR plates	Applied Biosystems
cell culture 96-and 24-well plates	Sarstedt
cell culture flasks (250 ml)	Sarstedt
cell culture plates (5 & 10 ml adherent and suspension cells)	Sarstedt
cell strainer	Becton Dickinson GmbH
cryo tubes	Nunc A/S
FACS tubes	Becton Dickinson GmbH and Sarstedt
nitrocellulose membrane 200 mm × 200 mm	Schleicher & Schuell
pasteur pipettes	Wilhelm Ulbrich Mainz
pipette tips (200 μl & 1000 μl)	Sarstedt
sterile bottle top filter 500 ml	Nalgene
sterile pipettes 10 ml	Sarstedt
sterile filters pore size 0.2 μm	Sarstedt
syringes 10 ml	Dispomed Witt oHG
weighing paper MN 26	Macherey-Nagel GmbH & Co. KG
Whatman paper GB 003	Schleicher & Schuell

3.10 Computational analysis

The computational analysis of ^{51}Chromium and [^{3}H]-Thymidine assays was performed using MicroBeta software for the Wallac MicroBeta Trilux liquid scintillation counter and Excel 2003 (Microsoft). Pictures of immunoblots developed with ECL where analysed with Chemilux and densitometric analysis was done with gel pro analyzer 4.5 from Media Cybernetics. The analysis of fluorescence activated cell sorting (FACS) data was performed using Cell Quest (BD Biosciences) and subsequent mathematical and visual analysis was conducted with Excel 2003 or Weasel v2.5 for linux. RNA quality analysed on RNA 6000 Pico Chips were analysed with the BioAnalyzer 2100 software. The microarray data were read with G2567AA feature extraction software V.8.1 from Agilent Technologies and the analysis was made with algorithms for microarray analysis in R. The qRT-PCR with the ABI 7500 SDS software. The analysis of variances (ANOVA) was conducted with SPSS version 15.00, the conservative adjustment of Bonferroni and a type I error rate (α) set to 0.05. Wilcoxon-tests, Kruskal-Wallis-tests (H-tests), and Mann-Whitney-tests (U-tests) were performed with WinSTAT for Excel 2003 and the type I error rate α was always set to 0.05 unless otherwise stated.

3.11 List of providers

Addresses of providers mentioned in this material part are listed in table 3.14.

Table 3.14: List of providers

Name	Address
Agilent Technologies Deutschland GmbH	Herrenberger Str. 130, 71034 Böblingen
Amersham Biosciences	Munziger Str. 9, 79111 Freiburg
AppliChem GmbH	Ottoweg 4, 64291 Darmstadt
Applied Biosystems	Frankfurter Str. 129b, 64293 Darmstadt
Ares Bioscience GmbH	Graeffstr. 5, 50823 Köln
Axxora	Gallusstr. 10, 35305 Grünberg
Bachem Distribution Services GmbH	Hegenheimer Str. 5, 79576 Weil am Rhein
Bandelin electronic	Heinrichstr. 3-4, 12207 Berlin
Becton Dickinson GmbH	Tullastr. 8-12, 69126 Heidelberg
Bethesda Research Laboratories GmbH	Postfach 210265, 7500 Karlsruhe 21
Biochrom AG	Leonorenstr. 2-6, 12247 Berlin
Biomed Labordiagnostik GmbH	Bruckmannring 32, 85764 Oberschleißheim
Biometra biomedizinische Analytik GmbH	Rudolf-Wissell-Str. 30, 37079 Göttingen
Biomol/Stressgen	Waidmannstr. 35, 22769 Hamburg
Biotec-Fischer GmbH	Daimlerstr. 6, 35447 Reiskirchen
Brand GmbH + Co KG	Otto-Schott-Strasse 25, 97877 Wertheim
Caltag Laboratories	Brauhausstieg 15-17, 22041 Hamburg
Carl Zeiss MicroImaging GmbH	Königsallee 9-21, 37081 Göttingen

Continued on next page

3 Materials

Table 3.14 – continued from previous page

Name	Address
Clean Air Deutschland GmbH	Rubensweg 17, 40724 Hilden
Desaga (Sarstedt-Group)	In den Ziegelwiesen 1-7, 69168 Wiesloch
Dianova	Warburgstr. 45, 20354 Hamburg
Dispomed Witt oHG	Am Spielacker 10-12, 63571 Gelnhausen
Eppendorf AG	Barkhausenweg 1, 22339 Hamburg
Ernst Leitz Wetzlar GmbH	Now: Wild Leitz Holding AG
Genecraft	see: Ares Bioscience GmbH
GeReSo GmbH	Carl-Orff-Str. 33, 37574 Einbeck
Hamilton Bonaduz AG	Via Crusch 8, 7402 Bonaduz (Schweiz)
Hartmann Analytic GmbH	Inhoffenstr. 7, 38124 Braunschweig
Heraeus Holding GmbH	Heraeusstraße 12-14, 63450 Hanauy
ICN/ MP Biomedicals	Thüringer Str. 15, 37269 Eschwege
IKA Werke GmbH & Co. KG	Janke & Kunkel-Str. 10, 79219 Staufen
Intas Science Imaging Instruments	Florenz Sartorius Str. 14, 37079 Göttingen
Invitrogen	Emmy-Noether-Str. 10, 76131 Karlsruhe
LKB Instruments	Now: Amersham Biosciences
Macherey-Nagel GmbH & Co. KG	Neumann Neander Str. 6-8, 52355 Düren
Merck Biosciences GmbH	Ober der Roeth 4, 65824 Schwalbach/Ts
Microsoft Deutschland GmbH	Konrad-Zuse-Straße 1, 85716 Unterschleißheim
Miltenyi Biotec	Friedrich-Ebert-St. 68, 51429 Bergisch Gladbach
Nalgene	see Nunc A/S
New England Biolabs GmbH	Brüningstr. 50, 65926 Frankfurt am Main
Nunc A/S	Kamstrupvej 90, 4000 Roskilde, Denmark
Omnilab-Krannich GmbH	Elliehäuser Weg 17, 37079 Göttingen
PerkinElmer LAS GmbH	Ferdinand-Porsche-Ring 17, 63110 Rodgau-Jügesheim
Peqlab Biotechnologie GmbH	Carl-Thiersch-Str. 2b, 91052 Erlangen
Phase GmbH	Blücherstr. 2, 23564 Lübeck
Polaroid GmbH	Robert-Bosch-Str. 32, 63303 Dreieich-Sprendlingen
Promega GmbH	Schildkrötstr. 15, 68199 Mannheim
R & D Systems GmbH	Borsigstr. 7, 65205 Wiesbaden-Nordenstadt
Ratiopharm GmbH	Graf-Arco-Str. 3, 89079 Ulm

Continued on next page

3.11 List of providers

Table 3.14 – continued from previous page

Name	Address
Roche Diagnostics GmbH	Sandhofer Strasse 116, 68305 Mannheim
Roth GmbH & Co.	Schoemperlenstr. 1-5, 76185 Karlsruhe
PerkinElmer LAS GmbH	Ferdinand-Porsche-Ring 17, 63110 Rodgau-Jügesheim
Sarstedt AG & Co.	Postfach 1220, 51582 Nümbrecht
Sartorius AG	Weender Landstr. 94-108, 37075 Göttingen
Schleicher & Schuell GmbH	Hahnestr. 3, 37586 Dassel
Schott AG	Hattenbergstr. 10, 55122 Mainz
Schütt Labortechnik GmbH	Rudolf-Wissell-Str. 11, 37079 Göttingen
Serva Electrophoresis GmbH	Carl-Benz-Str. 7, 69115 Heidelberg
Sigma-Aldrich Chemie GmbH	Eschenstrasse 5, 82024 Taufkirchen
SPSS GmbH Software	Rosenheimer Str. 30, 81669 München
Thermo Scientific	Dieselstr. 4, 76227 Karlsruhe
VWR International GmbH	Hilpertstr. 20a, 64295 Darmstadt
Wallac	Now: PerkinElmer LAS GmbH
Wilhelm Ulbrich GdbR	Zum Eichelberg 10, 96050 Bamberg

4 Methods

4.1 Cell culture methods

4.1.1 Culturing of cells

Cell lines were maintained in $NaHCO_3$-buffered DMEM with 10 % FCS, 2 mM L-glutamine, 1 mM sodium pyruvate, 100 U/ml penicillin and 100 µg/ml streptomycin in dishes for tissue culture in a humidified atmosphere of 10 % CO_2 at 37 °C. For maintaining the cells in culture, adherent cells were detached with trypsin and splitted regularly to keep them in a logarithmic growth curve. Trypsin treatment was stopped with FCS-containing medium. Section 3.6 on page 31 describes all cell lines used. For storage, cell pellets were resuspended in 0.5 ml Hepes-buffered DMEM, 0.5 ml freezing medium was added, and the cell suspension was directly transferred into cryo tubes. After transfer, cryo tubes were immediately placed in a polystyrene box at -80 °C. For long-term storage, cryo tubes were placed into liquid nitrogen (-196 °C) after a few days. Cells were thawed by rinsing the cryo tubes under warm water until the pellet got fluid. The cells were then transferred into a 13 ml tube containing 10 ml Hepes-buffered DMEM and after centrifugation for 10 min at 300 accelaration of gravity (x g), the supernatant was discarded and cells were resuspended in $NaHCO_3$-buffered DMEM with all supplements and cultured under standard conditions.

4.1.2 Subcloning of the Wt clone of *Sulf* mouse embryonic fibroblasts by limiting dilution

To get Wt clones with a higher H2K^b expression than the parental Wt clone, this *Sulf* MEF clone was subcloned by limiting dilution. Wt cells were counted and resuspended in medium to get a concentration of 200 cells/100 µl. Three 96-well plates with 100 µl $NaHCO_3$-buffered DMEM were used per well and 100 µl of cell suspension were added per well to the first column of each plate. Cells were mixed by pipetting and 100 µl of each well were transferred into the wells of the next column. This procedure was repeated to the end of the plate and after mixing 100 µl of each well of the last column were discarded. If out of 8 wells in a column just 1/3 contains cells, they are considered to be clonal. After about a week and regular control by

4 Methods

microscopy, the expanded clones were transferred into wells of a 24-well plate. After about another week, the clones were analysed in flow cytometry by cell surface staining for $H2K^b$ expression. Two clones, E9 and F7, with a higher $H2K^b$ expression than the parental Wt clone were chosen for further experiments.

4.1.3 Acute induction of heat shock protein 70 in the human melanoma cell line Ge-tet

The transgenic rat *Hsp70-1* was induced in the human melanoma cell line Ge-tet with 5 µg/ml doxycycline for 24 hrs under the normal culturing conditions described above. Induction was confirmed afterwards by flow cytometry after intracellular staining using an antibody specific for inducible HSP70 (anti-HSP70) as described in 4.2.4.2 on page 47.

4.1.4 Induction of apoptosis using staurosporine

For induction of apoptosis in HSP70 overexpressing cells, Ge-tet cells were treated with 5 µg/ml doxycycline for 24 hrs followed by an addition of 1 µM Staurosporine for 20 hrs and plus 5 µg/ml doxycycline for the same time period.

4.1.5 Optimising multiplicity of infection for adenovirus

In initial experiments the transfection efficiency of different Ge cell lines and the Wt clone of the *Sulf* MEFs was determined by the addition of distinct amounts of AdV to the cells. This adjustment of the multiplicity of infection (MOI) of AdV was done to exclude variances in transfection efficiency. 4–7×10^4 cells per well were seeded into a 24-well plate. 24 hrs after addition of QBI-(AdV)-GFP at different MOI and incubation, the MOI was chosen as optimal MOI, in which 80–90 % of the cells were transfected by the AdV and expressed GFP as determined by flow cytometry. The calculation was done as follows. To 8×10^4 cells the AdV with a biological activity of 9×10^{10} pfu per ml was added. The amount of AdV to be added to the cells to get an MOI of 500 can be calculated by using the formula 4.1. Filling in the values leads to equation 4.2.

$$\text{virus } [\mu l] = \frac{1000 \; \mu l \times \text{MOI} \times \text{number of cells}}{\text{pfu}} \quad (4.1)$$

$$0.444 \; \mu l = \frac{1000 \; \mu l \times 500 \times 8 \times 10^4}{9 \times 10^{10}} \quad (4.2)$$

One would need 0.444 µl of the AdV. For better handling a 1:10 dilution of the AdV was made and 4.4 µl were used.

4.1.6 Induction of apoptosis using granzyme B and adenovirus

GrB is a protease, which is part of cytotoxic granules of CTLs and NK cells and together with the pore-forming protein perforin induces apoptosis in tumour cells. Attenuated AdV was used as endosomolytic agent (Froelich et al. 1996a) to deliver GrB into the cytosol of the human melanoma cell line Ge and MEFs. Depending on the experiment, the required number of wells of a 24-well plate were seeded with 4–10×10^4 cells per well. The next day, one well of each cell line was counted and the amount of AdV for optimal MOI was calculated (see formula 4.1 on the preceding page; Ge clones MOI 500; MEF MOI 500). The biological activity of the AdV-GFP was 9.5×10^{10} pfu per ml and the one of AdV-β-gal 9.6×10^{10} pfu per ml. Medium was replaced with DMEM with 1 % BSA per well and 1 ng/μl GrB was added to the respective wells. After 30 min QBI-AdV-GFP or AdV-β-gal were diluted and added to get a final volume of 200 μl per well. After 24 hrs supernatants as well as cells were harvested using 200 μl trypsin/EDTA/PBS for 5 min at room temperature (RT). Cells were washed off using PBS, transferred into 13 ml tubes and centrifuged for 5 min at 300 x g. This was followed by different tests to determine various steps in the apoptotic pathway.

4.1.7 Uptake of labelled granzyme B into cells with acute heat shock protein 70 overexpression and mouse embryonic fibroblasts

The internalisation of GrB was investigated by using fluorescence-labelled GrB. For MEFs, 10×10^4 cells were seeded per well of a 24-well plate. 6×10^4 Ge-tet or Ge-tra cells were seeded with 5 μg/ml doxycycline 24 hrs before the experiment. The acute overexpression of HSP70 was confirmed by flow cytometric analysis at the day of the experiment. After successful confirmation of HSP70 induction in Ge-tet cells or seeding and complete adherence of the MEFs to the 24-well plate, either 100 μl DMEM 1 % BSA with or without 1 ng/μl Alexa-488-labelled GrB was added to the wells. After an incubation time of 1 hour at 37 °C, cells were harvested and washed twice with trypsin/EDTA/PBS to remove GrB bound to the cell surface. Subsequently, cells were incubated for 10 min with 200 μl 1 nM PI/PBS for the exclusion of dead cells, which were PI positive. The uptake of labelled GrB was measured by flow cytometry.

4.2 Immunological methods

4.2.1 Generation of effector cells

Effector cells were antigen-specific stimulated cytotoxic T-lymphocytes (CTLs) or fresh or stimulated natural killer (NK) cells. They were used in ^{51}Chromium and [^{3}H]-Thymidine-release assays to induce cell death in target cells. Depending on whether human blood or spleens of

4 Methods

mice were used as source for the effector cells, and whether they should be antigen-specific, the isolation and stimulation of the cells was different.

4.2.1.1 Generation of SIINFEKL-specific cytotoxic T-lymphocytes from transgenic OT-I mice

For the generation of antigen-specific CTLs, the spleen of TCR-transgenic OT-I mice was dissected, and homogenised in 10 ml Hepes-buffered DMEM with a Tenbroeck homogeniser. After centrifugation for 10 min at 300 x g, cells were resuspended in 5 ml erythrocyte-lysis buffer and incubated for 3 min at RT. Incubation of cells in medium containing 8 ml supernatant from Con A-stimulated rat lymphocytes, 40 µl 1 nM β-mercaptoethanol, 4 µl recombinant murine IL-2 (20 µg/ml) and 4 µl SIINFEKL (10 µg/ml) in 32 ml DMEM leads to antigen-specific activation of SIINFEKL-specific $CD8^+$ cells after 4 to 5 days of incubation. For restimulation after 4 to 5 days, cells were washed again and stimulated with irradiated spleen cells from C57Bl/6 mice (30 Gy), which were treated with erythrocyte-lysis buffer as well, in addition to the stimulation with Con A supernatant, 2-mercaptoethanol, recombinant murine IL-2 and SIINFEKL, as mentioned above.

4.2.1.2 Generation of human natural killer cells from whole blood by density gradient centrifugation and negative MACS selection

Human NK cells were generated from blood of healthy donors, which was separated using a density gradient centrifugation, followed by a negative magnetic cell sorting (MACS) selection for NK cells. 100 ml blood, anti-coagulated with heparin, was used for the isolation of NK cells. For the density gradient centrifugation Biocoll was used, which allows separation of peripheral blood mononuclear cells (PBMCs) from the remaining blood due to the higher density of Biocoll in comparison to blood plasma (Noble and Cutts 1967). For the separation 4 ml of Biocoll were pipetted into a 13 ml tube and 8 ml of blood were slowly layered on top to keep two separate layers. After centrifugation for 20 min at 500 x g a white ring became visible at the border of the Biocoll and the plasma. The ring was harvested with a pasteur pipette and collected in a 50 ml tube. Following three washing steps with Hepes-buffered DMEM once for 10 min at 500 x g and twice for 15 min at 200 x g, cells were filtered through a nylon mesh and counted 1:10 diluted with 1 % acetic acid to destroy remaining erythrocytes and allow counting of nucleated PBMCs only. 1×10^8 cells were resuspended in 400 µl pre-cooled MACS buffer and 100 µl biotin-labelled antibody cocktail (containing antibodies against cell surface molecules not being present on NK cells: CD3, CD4, CD14, CD15, CD19, CD36, CD123, and CD235a) was added for 10 min at 4 °C. After the incubation time, 300 µl pre-cooled MACS buffer and 200 µl of anti-biotin microbeads were added for 15 min in the fridge. Following the second

4.2 Immunological methods

incubation step, the tube was filled with pre-cooled MACS buffer to a final volume of 10 ml and centrifuged for 10 min at 300 x g at 4 °C. Meanwhile, a LS-column was placed into the holder of the magnetic separator within the magnetic field, and pre-rinsed with 3 ml of MACS buffer. Supernatant of centrifuged cells was discarded, cells were resuspended in 500 µl cooled MACS buffer and directly applied to the column. NK cells were not labelled and therefore passed the column and were collected in a fresh 13 ml tube. The column was washed 3 times with 3 ml MACS buffer. The enriched NK cell fraction was counted again and the cells were either used freshly or stimulated for 4 days with 100 units/ml proleukin. A small fraction of enriched NK cells and of cells not separated by MACS were analysed in flow cytometry with the following antibodies to confirm the enrichment: CD3, CD4, CD8, CD16, CD56, and CD94.

4.2.2 Preparation of concanavalin A supernatants for the stimulation of effector cells

Con A supernatants were used for the generation of antigen-specific mouse CTLs, as they contain various cytokines. Con A is known for its mitogenic activity. Spleens from rats were dissected, homogenised in Hepes-buffered DMEM and centrifuged for 10 min at 300 x g. Supernatant was discarded and cells were resuspended in 40 ml Hepes-buffered DMEM with 5 % FCS and 200 µl Con A (1 mg/ml) per spleen for 4 hrs at 37 °C in a water bath in a cell culture flask. After this incubation, cells were harvested and centrifuged for 10 min at 300 x g. Supernatant was discarded, cells were resuspended in 20 ml per spleen of normal cell culture medium in cell culture flasks and incubated for 20 to 24 hrs at normal cell culture conditions. After incubation, cells were collected and centrifuged for 10 min at 1100 x g. Supernatants were aliquotted 8 ml each into 13 ml tubes and stored at -20 °C.

4.2.3 Cytotoxic assays

Cytotoxic assays were performed to determine the lytic or apoptotic effect of killer cells, i.e. CTLs and NK cells, on different target cells. The assays can be modified in such a way, that also diffferent ways of stimulation of killer cells can be compared or inhibitors or enhancers can be added and compared as well. The read-out of the assays is the measurement of the release of radio-labelled substances from the target cells into the medium upon lysis or apoptosis.

4.2.3.1 ^{51}Chromium release assay

Target cells were labelled by incubating 1×10^6 cells for 1 hr at 37 °C in 350 µl Hepes-buffered DMEM containing 120 µl FCS and 50 micro Curie (µCi) $Na_2^{51}CrO_4$ which binds to intracellular cytoplasmic proteins. After 1 hour cells were washed three times with Hepes-buffered DMEM

4 Methods

for 10 min at 300 x g. Experiment-specific modifications, e.g. adding of 0.25 µg/ml SIINFEKL peptide (Ovalbumin aa 257–264) or EGTA/MgCl$_2$, as an inhibitor of calcium-dependent killing, were made when necessary. Antigen-specific CTLs were pipetted into 96-well plates with 1×10^4 ^{51}chromium-labelled target cells at decreasing ratios of, e.g. 10:1 to 0.156:1, in triplicates. After centrifugation for 5 min at 20 x g to enable immediate contact between effector and target cells, the plates were incubated for 4 hrs at 37 °C in a final volume of 200 µl. Following the incubation, the microtiter plates were centrifuged for 1 min at 20 x g and 50 µl supernatant (1/4 of the total volume) were harvested. Chromium is found in the supernatant after destruction of the cell membrane (Brunner et al. 1968). 50 µl so-called "sediments" were harvested from each well after adding and mixing 5 µl of a non-ionic detergent solution containing 10 % Triton-X 100 to destroy the membrane in order to determine the total chromium count. Supernatants and sediments were transferred into 96-well scintillation plates and in order to measure the radioactivity 200 µl scintillator (Optiphase Supermix) were added to each well. The ^{51}Chromium release was measured with a MicroBeta Trilux counter. The lysis was calculated by using the formula 4.3 (Dressel et al. 2000, 2004b), whereby cpm are the counts per minute. As the cpms were determined for 1/4 of the total volume of the supernatant, the formular adjusts for that as well as for the 50 µl of sediments, which also just correspond to 1/3 of the total volume. The factor 100 is used in this formula to calculate the lysis in percentage.

$$[\%] \text{ lysis} = \frac{4 \times \text{cpm supernatant} \times 100}{3 \times \text{cpm sediments} + 1 \times \text{cpm supernatant}} \quad (4.3)$$

The specific lysis, which indicates the activity of cytotoxic cells, is calculated by subtracting the values of the spontaneous release, i.e. chromium release in the absence of effector cells, from the lysis values.

4.2.3.2 [^{3}H]-Thymidine release assay

Target cells were labelled with 5 µCi/ml [methyl-^{3}H]-Thymidine 20 to 24 hrs before the test to allow Thymidine uptake into newly synthesised DNA during mitotic cell division. The basic approach is the same as in the ^{51}Chromium release assay except for the fact that ^{51}Chromium release from proteins of the cytosol into the medium indicates the lysis of the cells, whereas [^{3}H]-Thymidine release from the DNA into cytoplasm or the medium indicates apoptotic DNA fragmentation. 24 hrs after labelling target cells with [^{3}H]-Thymidine, effector and target cells were counted and pipetted into 96-well plates in a final volume of 100 µl in decreasing ratios of, e.g. 10:1 to 0.625:1 in triplicates. After centrifugation for 5 min at 20 x g to enable contact between effector and target cells, the plates were incubated for 4 hrs at 37 °C. The harvesting procedure was already described in (Garner et al. 1994; Motyka et al. 2000). The pipetting scheme was the same in each 96-well plate with rows A to H and lanes 1 to 12. The serial

4.2 Immunological methods

dilution of effector cells was made in rows A to E, row F always contained labelled target cells in the absence of effector cells for the calculation of the spontaneous release and rows G and H also contained only labelled target cells in the absence of effector cells for the determination of the total Tritium count. For the measurement of counts per minutes (cpms) in the supernatant 100 µl Tris/EDTA/Triton X-100 pH 8.0 were added to each of the wells in rows A to F. The cell suspensions were transferred into 1.5 ml Eppendorf tubes, vortexed and centrifuged at 4 °C for 12 min at 21600 x g to get rid of intact nuclei containing the intact labelled DNA. For the total tritium count (= sediments) 100 µl 2 % SDS/NaOH were added per well to the rows G and H and the cells were lysed by vortexing. 50 µl of supernatant and 50 µl sediments were transferred into 96-well scintillation plates. Finally, 200 µl scintillator were added per well and the Tritium-release was measured with a MicroBeta Trilux counter. The low-energy β-radiation of tritium could be detected with liquid scintillation counting. From the cpms the specific [^{3}H]-Thymidine release was calculated with the formula 4.3 on the preceding page.

4.2.4 Flow cytometric analyses

Flow cytometry is a technique in which single cells can be examined. A laser excites fluorescent dyes to emit light, which is then measured by photo detectors. The presence of distinct proteins or receptors on or in the cells can be quantified using specific antibodies or dyes.

4.2.4.1 Cell surface stainings for flow cytometric analysis

For staining of cell surface molecules, adherent cells were either scratched off from the plates using a scraper or mildly detached using PBS/EDTA instead of trypsin to avoid degradation of cell surface molecules. After detachment, cells were transferred into FACS tubes and centrifuged for 5 min at 300 x g with PBS for washing. Supernatant was discarded and a specific antibody was added at 4 °C in a fridge for 30 to 60 min depending on its binding affinity. Afterwards, cells were washed with PBS and either directly analysed by flow cytometry or a secondary antibody was added, followed by a last washing step and subsequent flow cytometric measurement.

4.2.4.2 Intracellular flow cytometric analysis

For the analysis of intracellular proteins, cells needed to be fixed and permeabilised before staining. Depending on whether the primary antibody was directly labelled, an isotype control was used as a control and was added at the same time point to another tube. Then just two washing steps with saponin/PBS were required before and after the staining. If the primary antibody was not labelled and a secondary antibody was required to detect the primary one, then the secondary antibody alone was used as a control and added at the same time point

4 Methods

as the secondary antibody was added to the primary antibody. This procedure required three washing steps with saponin/PBS before, inbetween, and after the staining.

Per cell line 3 FACS tubes with 1×10^6 cells or less were used, washed for 5 min at 300 x g with PBS and subsequently fixed for 10 min at RT with 1 % PFA pH 7.2. One tube with cells was left unstained for the setup of the FACS flow cytometer instrument. The second tube either contained an isotype control or the secondary antibody only. To the third tube either a labelled primary antibody or a primary antibody and later on a labelled secondary antibody were added. Following the fixation with PFA, the cells were washed twice with PBS and subsequently with saponin/PBS for 5 to 10 min at 300 x g. Afterwards the specific antibody, e.g. anti-HSP70, was added to one of the tubes and incubated at RT for 45 to 60 min. After incubation, the cells were washed again with saponin/PBS and a secondary labelled antibody was added for 45 to 60 min at RT in the dark, e.g. goat anti-mouse IgG FITC-conjugated. Following the second incubation time, cells were washed with saponin/PBS and PBS again and were resuspended in a small amount of PBS for subsequent flow cytometric analysis.

4.2.4.3 DiD-staining of Ge cells for activation of caspase-3 after NK cell-induced apoptosis

To distinguish the activation of caspase-3 in target cells and in killer cells by flow cytometry, target cells were stained with the dye DiD. Firstly, Ge-tra and Ge-tet-1 cells were treated with doxycycline for 24 hrs and were harvested with PBS/EDTA for subsequent HSP70 expression and caspase-3 activation analysis by intracellular flow cytometry. 5×10^6 targets were stained in 1.5 ml Hepes-buffered DMEM with 7.5 μl of DiD Vybrant dye for 20 min at 37 °C. After the staining, cells were washed 3 times with pre-warmed Hepes-buffered DMEM for 5 min at 300 x g. 1.4×10^4 target cells (Ge-tra ±dox and Ge-tet-1 ±dox) were transferrred into 8 FACS tubes for a ratio of 5:1 and 0:1 killer to target cells. 7×10^5 NK cells were added to each tube with a ratio of 5:1. The tubes were filled to a final volume of 1 ml with Hepes-buffered DMEM 10 % FCS and were centrifuged for 2 min at 20 x g to enable cell contacts. Killer and target cells were co-incubated for 4 hrs at 37 °C for the induction of apoptosis. The subsequent intracellular flow cytometric analysis was performed as described in 4.2.4.2 on the preceding page. In the flow cytometric analysis, DiD positive cells were gated and the percentage of caspase-3 activation in DiD-labelled target cells was evaluated.

4.2.5 Measurement of apoptosis in cells

Many different methods exist to measure apoptosis in cells. The methods for analysing key steps are listed in the following, whereby most of them are based on flow cytometry using antibodies or dyes which stain certain proteins.

4.2 Immunological methods

4.2.5.1 Annexin V binding to phosphatidylserine on the cell surface

Annexin V is a 35 kDa protein, which can bind to PSs on the cell surface of apoptotic cells. The presence of PS, which translocates to the cell surface from the inner leaflet of the membrane is a hallmark of early apoptosis. FITC-labelled annexin V was used in flow cytometry to detect early apoptotic cells and in combination with PI to detect cells in a later stage of apoptosis, when the cell membranes are already leaky and PI can enter the cells. 2 μl annexin V-FITC were added into 250 μl annexin binding buffer and 0.8 μl of a 1:6.25 diluted PI solution was added for 20 to 30 min at 4 °C in the fridge to each tube. Cells were measured directly, unbound annexin was not washed off. Positive PI staining indicates necrotic or late apoptotic cells.

4.2.5.2 Release of cytochrome c from mitochondria

In the intrinsic apoptotic pathway, cytochrome c, a component of the electron transport chain, which is associated with the inner membrane of mitochondria, is released from mitochondria. Together with APAF1 and caspase-9 it forms an apoptosome, which can activate caspase-3, which leads to DNA fragmentation. The release of cytochrome c was measured by intracellular staining for flow cytometry (see section 4.2.4.2 on page 47) with an antibody specific for cytochrome c in the mitochondrial membrane (clone 7H8.2C12). In non-apoptotic cells, this antibody stains cytochrome c in the mitochondrial membrane. This staining is reduced upon release of cytochrome c in apoptotic cells as described (Stahnke et al. 2004).

4.2.5.3 Change in mitochondrial membrane potential

Another key event in the intrinsic apoptotic pathway is the loss of the mitochondrial membrane potential $\Delta\Psi$. The mitochondrial membrane potential $\Delta\Psi$ is a result of the asymmetric distribution of protons and other ions along the inner mitochondrial membrane. This unequal distribution of ions results in an electric ($\Delta\Psi$) and a chemical (ΔpH) gradient, both essential for the function of mitochondria. The inner leaflet of the inner mitochondrial membrane is negatively charged. The binding of the lipophilic cationic dye 5,5',6,6'-tetrachloro-1,1',3,3'-tetraethylbenzimidazolylcarbocyanine iodide (JC-1), which distributes in the mitochondrial matrix, correlates with the $\Delta\Psi$ (Cossarizza et al. 1993). JC-1 exists as a monomer and upon excitation at 490 nm it emits light at a wavelength of 527 nm (Hada et al. 1977). In non-apoptotic cells with a high $\Delta\Psi$, JC-1 forms J-aggregates, which emit at 590 nm (Reers et al. 1991). Thus, a loss of $\Delta\Psi$ is indicated by a shift in fluorescence from red (590 nm) to green (527 nm).

2.5 μg/ml JC-1 were diluted 1:1000 in prewarmed PBS 10 % FCS. 1 ml of diluted JC-1 was added to 3×10^5 cells for 5 min at 37 °C in an incubator without CO_2, followed by 3 washing steps with ice-cold PBS for 5 to 10 min at 300 x g. After addition of 60 μl PBS,

4 Methods

cells were analysed by flow cytometry. Living cells have JC-1 aggregates in their mitochondria and therefore show a high $\Delta\Psi$ with a red fluorescence, whereas apopototic cells have a loss in membrane potential resulting in a low $\Delta\Psi$ and JC-1 monomers with a green fluorescence.

4.2.5.4 Activation of caspase-8

Caspase-8 is an initiator caspase of the extrinsic apoptotic pathway. The proenzyme and active enzyme states of caspase-8 were detected with an antibody from clone 1C12. This antibody was used in immunoblots (see section 4.3.3 on page 52).

4.2.5.5 Activation of caspase-3

The activation of the effector caspase-3 is a central step in apoptosis, as the extrinsic and the intrinsic apoptotic pathways meet there. The activation of caspase-3 can be detected with an antibody from clone C92.605 as described in section 4.2.4.2 on page 47. The antibody exclusively detects the active but not the inactive form of caspase-3.

4.2.5.6 Sub G1-peak analysis to measure DNA loss

The sub G1-peak analysis is a flow cytometric proof of DNA fragmentation and chromatin condensation. The amount of DNA content in living cells depends on their stage in cell cycle. Cells in G0/G1-phase have a normal DNA content. During S-phase the content of DNA gets doubled and is finally doubled in G2-phase. The DNA content is halved afterwards during mitosis. Apoptotic cells can have a lower DNA content than cells in G0/G1-phase due to DNA fragmentation and staining is reduced due to chromatin condensation. By staining apoptotic cells with PI, which intercalates between bases of DNA, the DNA content can be visualised in flow cytometry. PI can only enter cells when the membrane is not intact and also stains RNA, so that RNase needs to be added to degrade RNA before the DNA content is analysed. After induction of apoptosis by various means, cells were resuspended in 500 μl PBS and pipetted into 10 ml ice-cold EtOH for fixation. After at least 18 hrs at -20 °C, tubes with EtOH were centrifuged for 10 min at 2000 rpm , EtOH was completely removed and pellets were washed with PBS for 10 min at 300 x g. Following discarding of PBS, cells were stained with 100–150 μl PBS/PI/RNase A for 30 min at 37 °C. Subsequently, a sub G1-peak measurement was conducted by flow cytometry to determine the percentage of apoptotic cells by placing a marker left of the G1-peak, which marks the percentage of apoptotic cells.

4.2.5.7 Apoptotic ladder to measure DNA fragmentation

One hallmark during the late stages of apoptosis is the fragmentation of DNA by DNases, such as CAD. This fragmentation is visible as an apoptotic ladder after electrophoretic separation.

Endonucleases can cut the chromosomal DNA, which is wound around a histone octamer, just inbetween two adjacent nucleosomes. The size of each rung of the apoptotic ladder therefore is a multiple of 180 bp, which equals the length of DNA wrapped in 1.67 left-handed superhelical turns around an histone octamer.

This method has been performed as described by Cossarizza et al. (1994). After induction of apoptosis by either heat-shock (30 min at 44 °C and 3 hrs at 37 °C for recovery), 1 μM staurosporine, or 1 ng/μl GrB plus AdV, cells were harvested, washed once in PBS for 10 min at 300 x g, transferred into 1.5 ml tubes and washed again with PBS for 10 min at 300 x g. The supernatant was sucked away using a pipette and the pellet was resuspended in 20 μl lysis buffer for DNA fragmentation. These reactions were incubated for 1 hr at 50 °C in a water bath for lysis of cells and degradation of proteins. To remove RNA, 25 mg/ml RNase A was added for 1 hr at 50 °C. To inactivate the enzymes, the temperature was increased to 70 °C for 2 min. After the addition of 10× DNA loading dye, the samples were loaded onto an 2 % TPE-agarose gel and separated overnight at 15 V. The apoptotic ladder was detected by UV-light, which visualised the ethidium bromide in the gel bound to the DNA.

4.3 Biochemical methods

4.3.1 Preparation of cell lysates for immunoblot analysis

For the preparation of lysates for immunoblots cells were harvested by detaching them without trypsin or if cell surface molecules on adherent cells were analysed, the cells were scratched off. Cells were counted using a Neubauer counting chamber and washed 4 times with PBS. After the last washing step, the remaining liquid was removed by pipetting and cell pellets were frozen at -20 °C. 15 μl reducing sample buffer (see table 3.7 on page 27) were added per 1×10^5 cells, the samples were boiled for 5 min at 95 °C in a water bath, shortly vortexed afterwards and centrifuged for 5 min at 10 000 x g to get rid of cell debris. Usually 40 μl of each sample were loaded onto a SDS gel. For the lysates for the CAR analysis, it was essential that the cells were seeded in the same densities as the expression of HSPGs were described to vary with cell density (Dechecchi et al. 2001).

4.3.2 SDS-PAGE

One way of separating proteins is according to their size and charge. In the polyacrylamide gel electrophoresis (PAGE) (Laemmli 1970) in the presence of SDS secondary and non-disulfide-linked tertiary structures of the native protein are destroyed. The anionic detergent SDS adds negative charge to the protein. To reduce disulfide bridges β-mercaptoethanol is added. SDS binds uniformly to all proteins conferring a mass:charge ratio, which is proportional to the size

4 Methods

of the protein. Upon application of an electric field proteins are then separated according to their size. Polyacrylamide is used, as it is thermo-stable, chemically relatively inert, easy to polymerise, and the pore size can be modified by the amount of TEMED added.

All buffers and solutions necessary for SDS-PAGE and immunoblot analysis are listed in table 3.7 on page 27. The separating gel was poured into a chamber consisting of two glass plates, which were separated by spacers and fixed with clamps. After polymerisation a stacking gel was poured on top, into which a comb was inserted for forming of pockets. After polymerisation of the stacking gel, the gel was fixed in a vertical electrophoresis chamber, both buffer reservoirs were filled with running buffer and air bubbles were removed. The lysates were loaded onto the gel with a Hamilton syringe. The electrophoretic separation was performed at 40 mA until the dye front reached the bottom of the gel.

4.3.3 Immunoblot

Immunoblot is a technique to transfer proteins after a SDS-PAGE from the polyacrylamide gel onto a nitrocellulose membrane by applying an electric field (Towbin et al. 1979). The proteins are fixed on the membrane like a copy of the gel and are freely accessible by antibodies.

After the SDS-PAGE, the gel was placed into transfer buffer for equilibration. The blot was set up in the following order, whereby except for the outermost perforated plastic plates, everything else was soaked in transfer buffer: perforated plate, sponge, blotting paper, gel, nitrocellulose membrane, blotting paper, sponge, perforated plate. The stack was placed into a vertical chamber at 4 °C containing transfer buffer. Perpendicular to the gel an electric field was applied so that the negatively charged proteins in the gel were transferred onto the nitrocellulose membrane. It was therefore important, that no air bubbles are between the gel and the nitrocellulose membrane, when the stack is set up. Transfer was done overnight at 16 V.

After transfer, the gel was removed from the stack and equilibriated in PBS/Tween. To determine whether the blotting procedure was successful, the blot was stained with Ponceau S, an azo dye, which binds reversibly to the positively charged amino groups of proteins. Protein lanes and bands of a length standard were marked with a ball-pen and the blot was cut to a proper size. Depending on the antibodies used for developing the immunoblot, the blot was cut into an upper and a lower part for development with two different antibodies, or left as one blot for development with just one antibody. The addition of antibodies to the blot was always followed by washing off unbound antibodies with PBS/Tween. The primary antibody was either added for 2 hrs at RT in PBS/Tween or overnight at 4 °C. Afterwards, blots were washed 3 times for 5 to 10 min with PBS/Tween on a shaker before the secondary antibody was added for 60 min at RT. This second incubation time was followed by 3 washing steps for at least 10 min each with PBS/Tween.

The blots were either developed with DAB or with the more sensitive method ECL. The addition of reduced DAB and H$_2$O$_2$ for developing the immunoblot leads to an oxidation of DAB and the formation of water by the enzyme peroxidase (HRP), which is conjugated to the secondary antibody (see equation 4.4). The oxidation of DAB results in a brown colour at the site where the peroxidase converts the DAB, which is then visible on the blot.

$$\text{DAB}_{\text{red}} + \text{H}_2\text{O}_2 + 2\text{H}^+ \xrightarrow{\text{peroxidase}} \text{DAB}_{\text{ox}} + 2\text{H}_2\text{O} \qquad (4.4)$$

The development of the immunoblot with ECL was done by addition of luminol and H$_2$O$_2$, which are oxidised by the HRP to an excited form of 3-aminophthalate (3-APA) (see equation 4.5). The decay of the excited form of 3-APA to a lower energy state is responsible for the emission of light. With this method proteins can be visualised even, if they are present in femtomole quantities and the light signal can be captured by exposing the blot to the camera of an imaging system (Chemilux system from Intas).

$$\text{luminol} + \text{H}_2\text{O}_2 \xrightarrow{\text{peroxidase}} 3-\text{APA}_{\text{excited}} \xrightarrow{\text{peroxidase}} 3-\text{APA} + \text{light} \qquad (4.5)$$

Primary antibodies were diluted as follows: 1:500 for CAR, 1:4000 for HSC70, 1:1000 for caspase-8. The secondary antibodies were conjugated with HRP and were diluted 1:10000 for ECL development and 1:4000 for DAB development.

4.3.4 Densitometric analysis of levels of coxsackie and adenovirus receptor on *Sulf* mouse embryonic fibroblasts

A densitometric analysis was performed, to compare levels of CAR with the internal loading control HSC70. This analysis allows to compare defined band intensities/volumes and depicts them as an integral of optical density. The internal control, in our case HSC70, was set to 100 % so that the ratios of the integrals of optical density for CAR could be calculated. The programm used was gel pro analyzer 4.5 from Media Cybernetics.

4.4 Molecular biological methods

4.4.1 RNA isolation from cells

For analysing the transcriptome of cells, RNA was isolated. In comparison to DNA, RNA needs to be handled carefully as it is prone to degradation. Therefore, all materials used for RNA isolation must be RNase-free and RNA requires to be cooled on ice all the time.

4 Methods

4.4.1.1 RNA isolation

This protocol can be used to prepare very small amounts of RNA down to 100 µg total RNA or less (Novota et al. 2008). Cell pellets were either snap-frozen in liquid nitrogen and stored at -80 °C or directly resuspended in 360 µl Trizol, and incubated at RT for 5 min. After centrifugation for 10 min at 4 °C at maximum speed, the upper liquid phase was transferred into a new tube, 72 µl chloroform were added and the tubes were vortexed for 30 seconds. The tubes were centrifuged again for 5 min at 4 °C at 12300 x g and the uppermost clear aqueous phase was transferred into a clean 1.5 ml tube. For precipitation of nucleic acids 0.8 volumes of isopropanol were added to the clear aqueous phase, shaked and mixed by vortexing for 10 seconds. The tubes were then incubated for 30 min at -80 °C. After precipitation, the hand-warmed samples were shortly vortexed, and centrifuged for 30 min at 4 °C at maximum speed. The supernatant was discarded and the pellets were washed with 70 % EtOH for 10 min at 4 °C at maximum speed. Supernatant was discarded again. The pellets were air-dried at RT and redissolved in 50 µl RNase-free H_2O.

In the next step, the isolated RNA was purified and contaminating traces of DNA were removed. To each tube, the following substances were added to destroy DNA with DNases on one hand and to protect the isolated RNA from RNases on the other hand: 42.5 µl RNase-free H_2O, 5 µl 1 M Tris/HCl pH 7.5, 1 µl 1 M $MgCl_2$, 1 µl RNase-free DNase I (1 U/µl), 0.5 µl RNase inhibitor (40 U/µl). The tubes were briefly vortexed and incubated for 20 min at 37 °C in a waterbath. Afterwards, 1 volume of phenol/chloroform/isoamylalcohol (25/24/1 (v/v/v)) was added per tube of DNase I treated RNA and was briefly mixed by vortexing. The tubes were centrifuged for 2 min at RT at 12300 x g. After centrifugation, the upper liquid phase containing the RNA was transferred into new 1.5 ml tubes and 1 volume of isopropanol and 0.1 volume of 3 M sodium acetate pH 4.8 (final concentration 300 mM) were added and tubes were briefly vortexed. The RNA was precipitated for 15 min at -80 °C and after hand-warming and vortexing, the tubes were centrifuged at 12300 x g for 15 min at 4 °C. The supernatant was discarded and 1 ml of 70 % EtOH was added per tube for washing for 5 min at 12300 x g at RT. The washing step was repeated once and all liquid was removed to air-dry the pellet at RT. Pellets were resuspended in 20 µl RNase-free H_2O and left on ice for 15 min followed by mixing using a vortexer to allow RNA to get dissolved. A few microliter of RNA were taken out of each tube for measuring the concentration and quality.

4.4.1.2 Determining RNA concentration

The concentration of isolated RNA from cells was measured against a blank of H_2O using the NanoDrop 1000 spectrophotometer and the appropriate software NanoDrop 1000 3.3. For this purpose the blank was set to zero and then 1 µl of sample RNA was added, sucked in and

4.4 Molecular biological methods

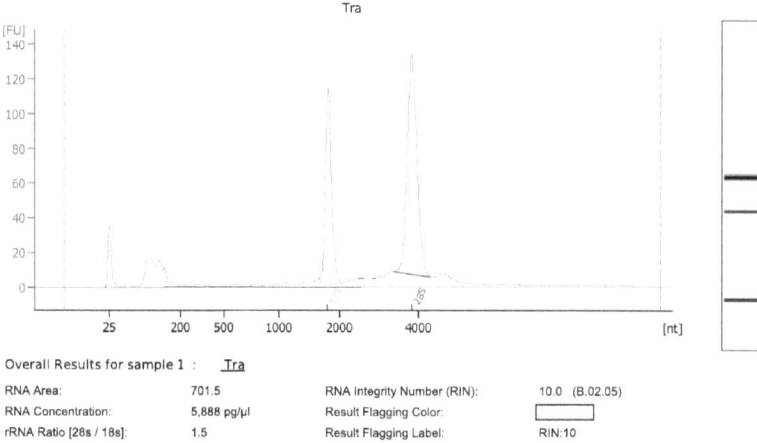

Figure 4.1: **RNA quality analysis with an RNA 6000 Pico Chip** Depicted is one example of an electropherogram from one RNA preparation. In the electropherogram the loaded length marker with a size of 25 nucleotides and the 18S and the 28S ribosomal subunit peaks are visible. The RNA integrity number (RIN) of this sample was 10.0, showing a complete intact RNA. The x-axis gives the size of the RNA fragments in nucleotides (nt) and the y-axis represents fluorescence units (FU).

measured. The absorbance at a wavelength of 230 nm is given and calculated directly into a concentration given in ng/μl. Typical yields of RNA from 5 cm plates seeded with different Ge clones for RNA isolation were around 1 μg/μl, if the pellet was resuspended in 20 μl after isolation. If the concentration was a lot higher, RNA was further diluted with RNase-free H_2O before storage at -80 °C for better handling.

4.4.1.3 Determining RNA quality by RNA 6000 Pico Chip analysis

After isolating RNA the quality was determined to decide whether the RNA was suitable for microarray analysis and qRT-PCR. For assessing the quality, RNA 6000 Pico Chip kits from Agilent were used together with an Agilent 2100 BioAnalyzer based on the principle of capillary gel electrophoresis. RNA is electrophoretically separated on microfabricated chips and analysed by laser induced fluorescence detection. The BioAnalyzer software generates an electropherogram. An example of an electropherogram is shown in figure 4.1.

Specific algorithms in the programme take into account the whole RNA trace, not just the ratio of 18S to 28S ribosomal subunits. The state of RNA degradation is given as RIN, which is a value between 1 and 10, whereby 1 is the most degraded profile and 10 the most intact. The classification of RNA quality into numbers from 1 to 10 allows comparison between different samples (Mueller et al. 2004).

4.4.2 Microarray analysis

The setup of the microarray experiment was as depicted in figure 4.2 with Ge-tra and Ge-tet-1 both untreated and treated with doxycycline for 24 hrs. Technical replicates and dye-swaps were made to ensure high reliability of the results.

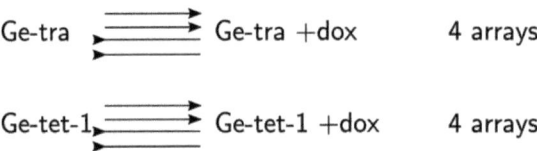

Figure 4.2: Setup of the microarray The control clone Ge-tra was compared with the doxycycline treated Ge-tra and the inducible clone Ge-tet-1 with the clone Ge-tet-1 acutely overexpressing HSP70 after doxycycline treatment for 24 hrs. Both comparisons were made in technical replicates and with dye swaps, meaning that in 2 out of 4 arrays the untreated sample was labelled with Cy3 and in the other 2 the treated samples. In total 8 whole human genome microarrays with 44000 60-mer oligos were used.

Isolated RNAs from Ge-tra and Ge-tet-1 cells from three different treatments with doxycycline were pooled in equal amounts (500 ng each). As the RNA integrity was already determined, the first step of the oligo two-colour microarray was to amplify and to label the target RNA to generate complementary cRNA. This was achieved in two sequential reactions. In the first reaction, unlabelled dsDNA is generated from mRNA primed with an oligo (d)T-T7 promoter primer by reverse transcription using MMLV-reverse transcriptase without amplification. In the second reaction, Cy3 or Cy5-labelled amplified single-stranded cRNA is generated by a T7 RNA polymerase using an anti-sense promoter as depicted in 4.3 on the facing page.

For the preparation of the transcription of mRNA into dsDNA 2 μl of each 1:3200 diluted Spike-mix A or B were denatured together with 8.3 μl RNA (1500 ng) and 1.2 μl of T7 promoter primer for 10 min at 65 °C in a thermocycler. The reactions were placed on ice afterwards for 5 min, while the cDNA master mix was prepared (table 4.1). The addition of Spike-mix A and B serves the purpose of an internal quality control. They contain 10 different sequences of AdV in different concentrations, which bind to distinct dots on the microarray. On the basis of their fluorescence intensities and ratios between each other, it can be deduced whether the labelling and the hybridisation worked optimally for all samples.

Table 4.1: Master mix for dsDNA for microarray To generate dsDNA from mRNA.

Master mix for the generation of dsDNA per reaction

4 μl	5 × First strand buffer
2 μl	0.1 M DTT
1 μl	10 mM dNTP mix
1 μl	MMLV-reverse transcriptase
0.5 μl	RNaseOUT

4.4 Molecular biological methods

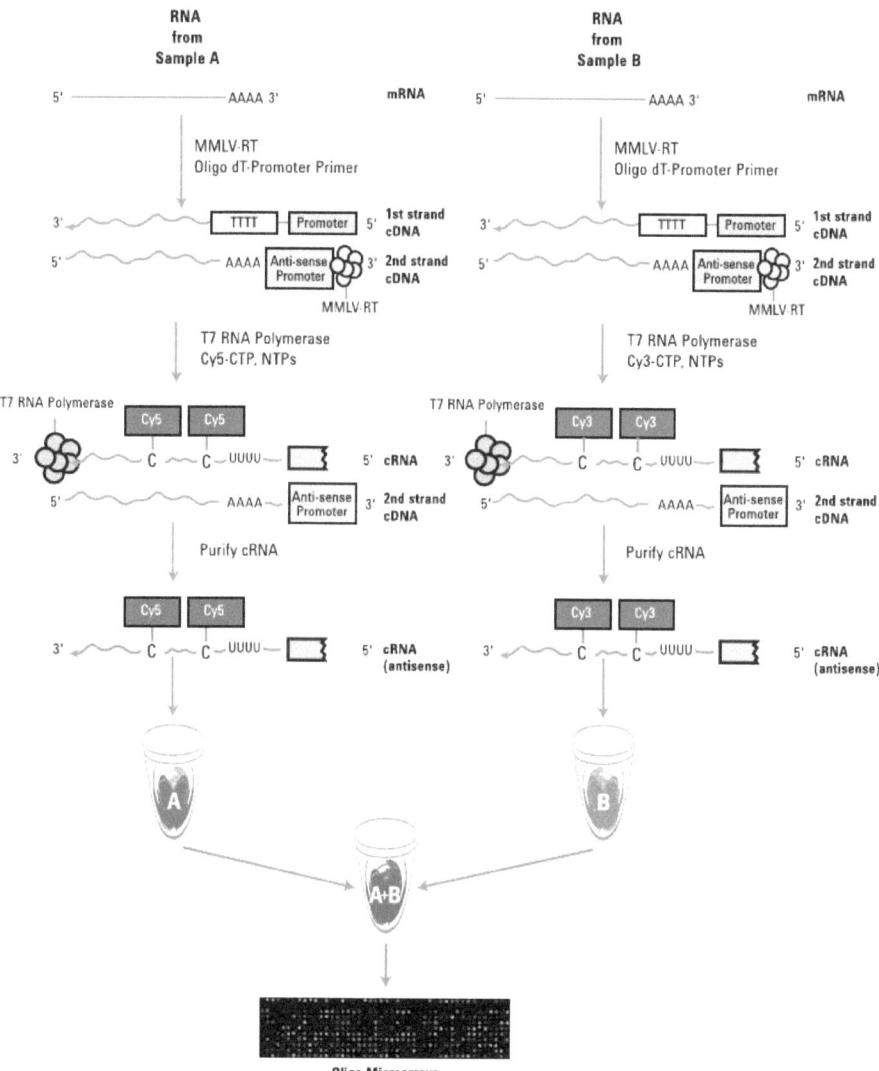

Figure 4.3: Generation of labelled cRNA for a two-colour microarray The scheme shows the amplification of mRNA into double-stranded cDNA and from this the generation of either Cy3 or Cy5-labelled cRNA. The scheme is taken from the Agilent technologies protocol "Two-Color Microarray-Based Gene Expression Analysis (Quick Amp Labeling)" version 5.7 from March 2008.

4 Methods

The denatured reactions and the cDNA master mix were mixed and all samples were incubated in a thermocycler. First the samples were incubated for 2 hrs at 40 °C and then for 15 min at 65 °C, before placing them for 5 min on ice. The generation of dsDNA was followed by the generation of amplified labelled cRNA by T7 RNA polymerase. To achieve that, each dsDNA sample was mixed with 60 μl of transcription master mix (table 4.2) and incubated for 2 hrs at 40 °C in a thermocycler.

Table 4.2: Master mix for generating labelled cRNA for microarray To be able to distinguish the two different cRNAs, added to the array later on, they need to be labelled with diferrent dyes, here Cyanine 3-CTP or Cyanine 5-CTP.

Master mix per 60 μl reaction to generate labelled cRNA	
20 μl	4× Transcription buffer
6 μl	0.1 M DTT
8 μl	NTP mix
6.4 μl	50 % PEG
0.6 μl	Inorganic pyrophosphatase
0.8 μl	T7 RNA polymerase
2.4 μl	Cyanine 3-CTP or Cyanine 5-CTP
0.5 μl	RNaseOUT
Ad 60 μl	RNase-free H$_2$O

After producing amplified labelled cRNA, the next step was to purify and to quantify it, in order to apply the different cRNA samples in the right amount to the array. For the purification of the amplified RNA Qiagen's RNeasy Mini kit was used. Briefly, the volume of each sample was filled to a final volume of 100 μl with RNase-free water, 350 μl RLT buffer were added and well mixed. 250 μl 100% EtOH were added to the tubes and mixed by pipetting up and down to precipitate the RNA. Samples were transferred to RNeasy mini columns, centrifuged for 30 sec at 4 °C at 12300 x g and the through-flow was discarded. The columns were transferred into new collection tubes, and after washing the columns twice with 500 μl EtOH-containing RPE buffer for 30 sec and 1 min at 4 °C at 12300 x g, the columns were transferred into new 1.5 ml tubes. Finally, RNA was eluted by adding 30 μl RNase-free water directly onto the column, waiting for 1 min, followed by centrifugation for 30 sec at 4 °C at 12300 x g.

This purified RNA was then quantified using NanoDrop ND-1000. The exact yield of labelled RNA per sample was calculated with the measured concentration of cRNA using formula 4.6, whereby the elution volume was 30 μl.

$$\mu\text{g cRNA yield} = \frac{(\text{concentration of cRNA}) \times (\text{elution volume})}{1000} \tag{4.6}$$

With the measured values for the concentration of Cy3 and Cy5 and formula 4.7 the specific activity defined as pmol dye per μg cRNA can be calculated.

$$\text{specific activity [pmole}/\mu\text{g]} = \frac{\text{concentration of Cy3 or Cy5 [pmole}/\mu\text{l]}}{\text{concentration of cRNA [ng}/\mu\text{l]} \times 1000} \quad (4.7)$$

A yield of less than 825 ng cRNA per array and a specific activity of less than 8.0 pmole/μg is not sufficient and the cRNA preparation needs to be repeated in order to proceed. This was not the case for any of our samples (see table B.1 on page 175).

The next step was to prepare the 10× blocking agent and the hybridisation samples for the microarray. The blocking agent was solubilised by adding 500 μl RNase-free H$_2$O to a tube containing lyophilised 10× blocking agent supplied with the Agilent Gene Expression Hybridization Kit and by vortexing gently. The sizes of the cRNA hybridisation samples for the microarray ranged between 50 and 3000 bases and were fragmented into shorter ones with an optimal size of 50 to 200 bases to improve target specificity to the 60mer oligos of the microarray. Hybridisation samples were fragmented by exposure to zinc acetate, which is the main component of the fragmentation mix given in table 4.3. cRNA was mixed with the fragmentation mix in a 1.5 ml tube and placed in the pre-warmed hybridisation oven for exactly 30 min at 60 °C. The reaction was quenched by carefully mixing the fragmented cRNA with 55 μl 2× GE hybridisation buffer Hi-RPM, which contains an excess of EDTA.

Table 4.3: Fragmentation mix for one 4× 44K microarray For the fragmentation of the labelled cRNAs, the following components needed to be mixed for each one of the four arrays. Either Cy3 or Cy5-labelled cRNA was used in the different fragmentation mixes.

55 μl fragmentation mix for one 4× 44K microarray	
825 ng	Cy3-labelled cRNA
825 ng	Cy5-labelled cRNA
11 μl	10× blocking agent
2.2 μl	25× fragmentation buffer
ad 50 μl	RNase-free H$_2$O

The hybridisation assembly was performed in an ozone-free room and nitrile gloves were worn during working with the arrays. Clean gasket slides (backing) were loaded into Agilent SureHyb chamber bases with the labels facing up. The hybridisation samples, 100 μl each, were slowly dispensed onto the backing without touching its walls. Two arrays with the "active side" down were carefully placed onto the backings, so that the numeric barcodes were facing up. Afterwards, the SureHyb chamber covers were placed onto the sandwiched slides and the clamps were handtighten onto the chambers. The assembled chambers were rotated vertically to wet the backing and to assure the mobility of air bubbles. The slide chambers were placed for 17 hrs into the hybridisation oven set to 65 °C at 10 rounds per minute (rpm).

After 17 hrs of hybridisation, the slides were washed with two different wash solutions. 1000

4 Methods

ml wash solution 1 containing 6× SSPE, 0.005 % N-lauroylsarcosine in H_2O were used for 8 arrays and 500 ml of wash solution 2 with 0.06× SSPE, 0.005 % N-lauroylsarcosine in H_2O were used. A series of washing steps was prepared: firstly 250 ml wash solution 1, secondly 250 ml wash solution 1 on a stirrer, thirdly 250 ml wash solution 2 on a stirrer, followed by 250 ml acetonitrile. Hybridisation chambers were dissembled in washing solution 1 and "active sites" were collected in a slide rack in wash solution 1 on a stirrer. The analysis of the microarray was done by scanning the arrays with a resolution of 5 μm and two different laser intensities with a scanner from Agilent Techonologies. First, the laser scanned the arrays with 100 % intensity and then again with 10 % intensity to get reliable data of all spots, which were overamplified with 100 % laser intensity. The data of the scanned arrays were then exported with the software accompanying the scanner and distinct algorithms were used in R to normalise the data and to calculate the logarithmic expression data and p-values.

4.4.3 Quantitative real-time PCR

The method of qRT-PCR is based on conventional PCR (Saiki et al. 1988). The principle is simple: Double stranded template DNA is denatured at 94 °C and cooled down to 55 °C to allow oligonucleotide primers to bind to the single-stranded DNA (annealing). Then, the samples are heated to 72 °C, the optimal working temperature for the Taq-polymerase, so that the primers are elongated with nucleotides until a complete and exact copy of the double-stranded template DNA is present again. Per cycle, starting with the denaturation and ending with the elongation, the amount of DNA is theoretically doubled. Here, we used qRT-PCR, an advanced PCR method, to confirm the data gained in the microarray analysis. The dye SYBR green can intercalate into dsDNA, which increases its fluorescence. This increase in fluorescence caused by intercalation of SYBR green is measured after each cycle at the end of the elongation phase at 530 nm. As the amount of DNA theoretically doubles with each cycle, an exponential function should be visible after 40 cycles.

4.4.3.1 Transcription of RNA into cDNA with reverse-transcriptase PCR

For the qRT-PCR the sample RNA needed to be transcribed into cDNA. For this purpose, a PCR tube was filled to a final volume of 8 μl with H_2O after the addition of 2 μg isolated RNA and 1 μg random primers. The mix was denatured for 5 min at 70 °C to allow binding of primers to the RNA. Samples were snap-cooled on ice to stop the reaction. The master mix for the reverse transcription with random primers is depicted in table 4.4 on the facing page. Master mix and denatured RNA were mixed and tubes were incubated at 37 °C for 60 min for primer annealing and synthesis, followed by 85 °C for 5 min for inactivation of reverse transcriptase and cooling tubes down to 4 °C.

4.4 Molecular biological methods

Table 4.4: Master mix for reverse transcription of RNA To transribe RNA into cDNA the listed components need to be mixed. Given are the volumes per 25 µl reaction.

Master mix per 25 µl reaction	
5 µl	5× reverse transcriptase buffer
2 µl	10 mM dNTP mix
1 µl	0.1 M DTT
1 µl	RNasin plus RNase inhibitor (40 U/µl)
1 µl	MMLV-reverse transcriptase (200 U/µl)
ad 25 µl	RNase-free H$_2$O

4.4.3.2 Validating primers for quantitative real-time PCR with standard PCR

The design of the primers for qRT-PCR is pointed out in section 3.2 on page 23 and the sequences are shown in table 3.4 on page 23. Primers were solubilised upon arrival and stocks of 100 pmole/µl in HPLC-H$_2$O were frozen at -20 °C. Functionallity of the primers was tested by using a mixture of 1 µl of each of the generated cDNAs for the qRT-PCR. A mixed working dilution was made of all primers with a concentration of 5 pmole/µl each of the forward and the reverse primer. A standard reaction of 20 µl was pipetted for all primers on ice with all components mentioned in table 4.5. For each primer, a negative control was performed without cDNA to exclude contaminations. The PCR was run in a cycler with the programme described in table 4.6.

Table 4.5: Standard PCR reaction To validate the primers for their specificity, a standard PCR was performed with the listed components per 20 µl reaction.

Standard PCR reaction in 20 µl	
0.2 µl	Taq-polymerase (5 U/µl)
0.4 µl	dNTP mix (10 mM of each dNTP in stock)
2 µl	Primer mix (forward and reverse; 5 pmole/µl each)
2 µl	10× Taq-buffer
2 µl	DNA (10 ng/µl cDNA mix)
ad 20 µl	HPLC-H$_2$O

Table 4.6: Standard acPCR programme

Standard PCR cycler programme			
1.	94 °C	4 min	
2.	94 °C	15 sec	
3.	60 °C	25 sec (depends on primer sequence)	
4.	72 °C	35 sec (back to step 2.; 35 cycles)	
5.	72 °C	10 min	
6.	4 °C	forever	

After the PCR the complete samples were analysed by gel electrophoresis on a 2 % agarose-

4 Methods

TBE gel with ethidium bromide as described in section 4.4.3.3.

4.4.3.3 Agarose gel electrophoresis

Agarose gel electrophoresis is an efficient method to separate and identify DNA molecules of a length between 50 bp and 25 kilo base pairs (kb) by applying an electric field. An agarose gel is prepared by boiling 0.5 to 2 % agarose in an electrophoresis buffer like Tris/borate/EDTA (TBE) or Tris/acetate/EDTA (TAE) in the microwave. The smaller the DNA fragments, the higher should be the percentage of the agarose.

For validating the primers for qRT-PCR, which should result in amplicons with a size of 90 to 120 bp, 2 % agarose was dissolved by boiling in TBE-buffer. 0.1 μl ethidium bromide were added per ml of agarose solution before pouring the agarose into a caster and inserting the comb. The solidified gel was placed into a chamber filled with the same electrophoresis buffer used for the gel itself. 20 μl PCR samples were mixed with 2 μl 10× DNA loading dye to weigh the DNA down and prevent it floating out of the pockets. The addition of dyes like xylene cyanol or bromphenol blue to loading dyes allows an estimation of how far the DNA has already moved in the gel. In a 2 % agarose gel bromphenol blue migrates as fast as DNA molecules with a size of smaller than 100 bp and xylene cyanol as molecules of about 800 bp (Mülhardt 2006). For determining the approximate length of the PCR amplicons, a length standard with defined band sizes was loaded as well. An electric field of 130 V was applied so that the negatively charged DNA in the pockets of the gel moves through the gel towards the anode. For analysis of molecule size and for excluding contaminations in the negative controls, the gel was placed onto an UV-table and a picture was taken for protocolling the result. By comparison of the bands of the PCR products with the bands of the loaded length standard, the size of the PCR samples can be determined.

4.4.3.4 Quantitative real-time PCR using SYBR green

After transcribing RNA into cDNA and validating the primers, the actual qRT-PCR was performed. Here, human *GAPDH* was used as housekeeping gene to be able to compare the results of the RNA preparations with each other. The expression of *GAPDH* was not changed in both Ge clones with or without doxycycline as shown in the microarray analysis, and it therefore served the purpose for standardising the changes in expression in all genes analysed. The transcribed RNA of Ge-tra, Ge, Ge-tet-1 and Ge-tet-2 with and without doxycycline treatment were used with all primers shown in table 3.4 on page 23. The amplification reactions were carried out in volumes of 25 μl in 96-well plates for real-time PCR using Power SYBR Green PCR master mix in a ABI 7500 Real-Time-PCR System. All reactions were performed in triplicates with 10 ng of cDNA as template. A master mix for real-time PCR is shown in table 4.7 and

4.4 Molecular biological methods

the cycler programme in table 4.8.

Table 4.7: Master mix for quantitative real-time PCR

Master mix per 25 µl reaction	
12.5 µl	SYBR Green
2 µl	Primer mix (forward and reverse; 5 pmole/µl each)
4 µl	10 ng cDNA
ad 25 µl	HPLC-H_2O

Table 4.8: Quantitative real-time PCR programme

Real-time PCR cycler programme		
1.	50 °C	2 min
2.	95 °C	10 min
3.	95 °C	15 sec
4.	60 °C	60 sec (back to step 3.; 40 cycles)
5.	60 to 90 °C	1 °C per min (dissociation stage)

For quality assurance of the qRT-PCR, as SYBR green is not selective in binding, dissociation curves were made. For this, step 5 of the PCR programme listed in table 4.8 is responsible. It heats up all reactions stepwise by 1 °C per min to 90 °C after the end of the qRT-PCR. Optimally, all dissociation curves should overlap, as the amplicons are identical and therefore possess the same melting behaviour as depicted in figure A.1 on page 173. Contaminations, mispriming, or primer-dimer artefacts can be detected by looking at the dissociation curves. Furthermore, standard curves were generated for the calculation of the efficiencies of the different primers as described in section 4.4.3.5.

4.4.3.5 Evaluation of data using Pfaffl

The qRT-PCR data were analysed using ABI 7500 SDS software. Using the formula described by Weksberg et al. (2005), variations in the RNA concentration in different samples were normalised by correcting the cycle treshold (ct) values obtained in qRT-PCR for selected genes by the ct values obtained for *GAPDH* in the same sample. The relative changes of the expression of selected genes were calculated using the mathematical model for relative quantification of real-time PCR data described by Pfaffl (2001), which also takes into account variations of the amplification efficiencies of different primer pairs. The expression levels of selected genes in cells with acute HSP70 overexpression were compared with the levels in cells with normal HSP70 levels. Formula 4.8 on the next page depicts the Pfaffl equation with x as gene of interest and y as internal control (= *GAPDH*), whereby E is the efficiency of the primer pair of the respective

4 Methods

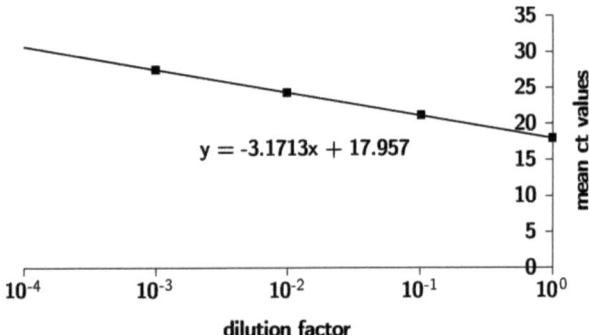

Figure 4.4: Calculation of the primer efficiency from the slope of cDNA dilutions for the *Hsp70* primer
The figure shows in the 4 mean ct-values of a mixed cDNA pool in 10-fold dilutions, starting with the undiluted pool with 10^0, of the *Hsp70* primer. The dilution 10^{-3} is a 1000-fold diluted cDNA pool of the original one. Through the 4 mean ct-values a straight line was put and its slope (-3.1713) gives the number, which needs to be put into equation 4.9 to get the efficieny of the primer.

sample:

$$\text{fold change in expression} = \frac{E_x^{(ct_x - ct_{x+dox})}}{E_y^{(ct_y - ct_{y+dox})}} \qquad (4.8)$$

The efficiency of the primer pair was determined by making serial dilutions of a pooled cDNA mix and adding the respective primer together with the SYBR green mix. A straight line is put through the mean ct values of the 4 different 10-fold dilutions, whereby the slope of the straight line is then put into the equation 4.9 for x and this then yields the efficiency of the primer pair, which was usually around 2. As an example for the calculation of efficiency from a slope, the efficiency of the *Hsp70* primer is shown in figure 4.4.

$$E = 10^{\frac{-1}{x}} \qquad (4.9)$$

4.4.4 Comparison of microarray and quantitative real-time PCR expression data

To be able to compare the microarray data with the data gained from the qRT-PCR, both data sets need to be modified. The easiest way to compare the data is to convert them to state the fold change in expression. The efficiency of the primers in qRT-PCR needs to be left out as it does not play a role in the microarray. The raw data of the microarray are $\log_2 x$ values, whereby x is the expression value of the respective gene in the untreated minus the treated sample. Thus, if the expression in the treated sample, here with doxycycline, is higher, the

4.4 Molecular biological methods

value is negative. To get the fold change in expression of the microarray data between untreated and treated samples (we do not have the single data, but just the result of the comparison) all data need to be converted with formula 4.10.

$$2^{-x} \qquad (4.10)$$

The raw data of the qRT-PCR, the ct values, need to be modified as well. This is done according to formula 4.11 with x being the gene of interest and y being the internal control ($= GAPDH$).

$$2^{((ct_x - ct_y) - (ct_{x+dox} - ct_{y+dox}))} \qquad (4.11)$$

The microarray data calculated with formula 4.10 can then easily be compared with the qRT-PCR data calculated with formula 4.11.

5 Results

5.1 Role of heat shock protein 70 in apopotosis

5.1.1 HSP70 overexpression in different Ge clones

The Tet-On system (Gossen et al. 1995) for acute overexpression of the stress-inducible HSP70 opens the opportunity to investigate effects of HSP70 expression in an unstressed cellular system (Dressel et al. 1999). The HSP70 expression in the control clone containing the transactivator domain is shown in part A of figure 5.2 on page 69. In contrast to the system for permanent overexpression of HSP70 (figure 5.1 on the following page), the HSP70 level is rather low in the system for acute overexpression (figure 5.2 on page 69). However, it could be shown earlier that acute but not permanent HSP70 overexpression increased the lysis of the Ge tumour cell lines by allogeneic CTLs (Dressel et al. 2003). Therefore, we wanted to analyse the molecular pathways, which lead to increased sensitivity to CTLs.

5.1.2 Gene expression analysis of cells acutely overexpressing heat shock protein 70

5.1.2.1 Whole human genome microarray analysis of Ge-tra and Ge-tet-1 cells

In a first attempt to determine whether *Hsp70* overexpression regulates the expression of other genes that might control susceptibility to CTLs, we performed a whole human genome microarray analysis.

For the microarray analysis doxycycline was added to Ge-tra and Ge-tet-1 cells for 24 hrs. Induction of HSP70 in Ge-tet-1 was validated by intracellular flow cytometry (see figure 5.3 on page 70) and RNA was isolated and pooled from 3 different doxycycline treatments. In addition to Ge-tra and Ge-tet-1 cells, parental Ge and Ge-tet-2 cells were used later for gene expression analyses by qRT-PCR. The microarray was performed as described in section 4.4.2 on page 56 and analysed as described in section 4.4.4 on page 64. A first visualisation of the microarray data of both clones was made via MA-plots, which show the logarithmic change in expression (M) on the y-axis and the mean logarithmic fluorescence intensity (A) on the x-axis (Dudoit et al. 2002). As it is assumed in gene expression analysis that most genes are not regulated,

5 Results

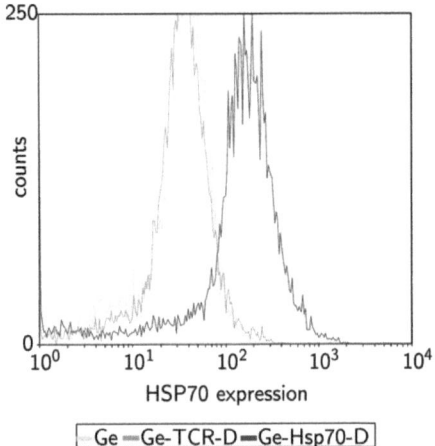

Figure 5.1: Permanent HSP70 overexpression in the Ge-Hsp70-D clone The intracellular expression of HSP70 was measured by flow cytometry with an antibody specific for HSP70 and a secondary antibody, which was FITC-conjugated. This is an overlay of three representative expression profiles of the parental cell line Ge, the control clone Ge-TCR-D overexpressing the β-chain of a TCR, and Ge-Hsp70-D, a clone permanently overexpressing HSP70.

the majority of the dots are usually located on the y-axis at zero. Part A of figure 5.4 on page 71 shows an MA-plot of the control clone Ge-tra, and part B depicts an MA-plot of the *Hsp70*-induced Ge-tet-1 clone.

Both MA-plots clearly show that only a few genes were regulated, as they were found outside of the grey cloud, which represents transcripts of genes with unchanged expression. The prominent change in expression of *Hsp70* itself in Ge-tet-1 cells was one indicator that the microarray gave reliable results. The *Hsp70* expression was induced (indicated by the circle) in Ge-tet-1 cells by the addition of doxycycline. In the control clone Ge-tra no *Hsp70* expression was induced upon addition of doxycycline as this clone just contained the transactivator domain.

We further analysed the number of genes up and down-regulated with a fold change in expression smaller than 0.25-fold or larger than 4.00-fold. The number of genes regulated in the respective clones are given in table 5.1 on page 70.

Bioinformatic analysis revealed that only 75 genes represented among the 44000 sequence tags on the array were regulated significantly above the indicated threshold by acute *Hsp70* overexpression in Ge-tet-1 cells and only 43 genes were regulated by doxycycline treatment in Ge-tra cells. Only two genes, *ASNS* and *CASP8* 8, were regulated in both clones by the addition of doxycycline.

In the 118 genes, which were significantly up- or down-regulated in at least one of both clones, genes involved in apoptosis were searched. The number of genes known to be involved in apop-

5.1 Role of heat shock protein 70 in apopotosis

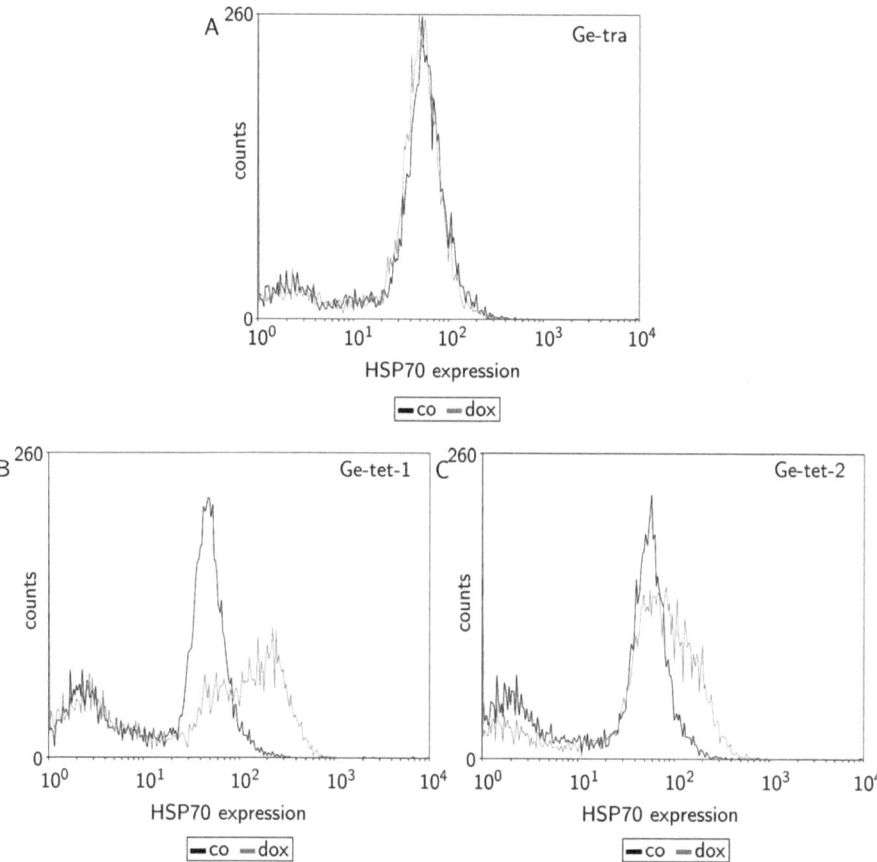

Figure 5.2: Acute HSP70 overexpression in Ge-tet-1 and 2 clones The intracellular expression of HSP70 was measured by flow cytometry with an antibody specific for HSP70 and a secondary antibody, which was FITC-conjugated. **A** A representative FACS histogram of intracellular HSP70 expression in Ge-tra cells without (co) and with doxycycline (dox) is shown. **B** A representative FACS histogram of intracellular HSP70 expression in Ge-tet-1 cells without (co) and with doxycycline (dox) is shown. **C** A representative FACS histogram of intracellular HSP70 expression in Ge-tet-2 cells without (co) and with doxycycline (dox) is shown.

tosis was limited. Thus, also other genes were chosen to validate the results of the microarray data by qRT-PCR experiments. The 13 selected genes including full name, expression values in Ge-tra in comparison to Ge-tra treated with doxycycline and Ge-tet-1 in comparison to Ge-tet-1 treated with doxycycline, and the respective p-values are shown in table 5.2 on page 72. A visualisation of the change in expression of the 13 genes upon doxycycline treatment in the microarray analysis is shown in part A of figure 5.5 on page 73. In this figure, it can be seen

5 Results

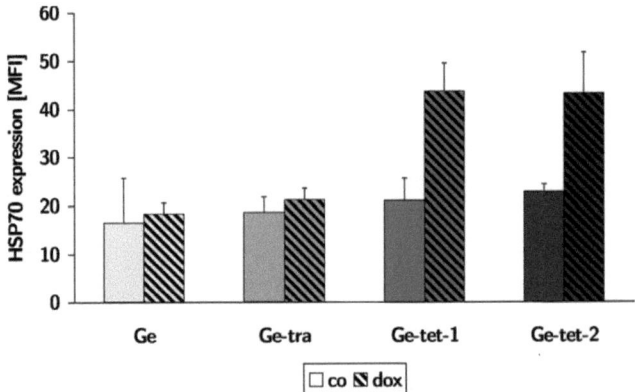

Figure 5.3: HSP70 expression in cells for RNA isolation Ge clones were treated with 5 μg/ml doxycycline (dox) for 24 hrs or were left untreated (co). HSP70 expression was determined by intracellular flow cytometry with an HSP70-specific antibody. Depicted are 3 independent doxycycline treatments of each clone, from which RNA was isolated for microarray analysis and qRT-PCR. Error bars depict standard deviations (SD).

Table 5.1: Analysis of microarray data This analysis depicts how many genes are up- and down-regulated in the control clone and the inducible one upon treatment with doxycycline.

number of genes regulated upon exposure to doxycycline	
total genes regulated in Ge-tet-1 ($Hsp70$ induced)	**75**
genes up-regulated in the $Hsp70$ overexpressing clone	60
genes down-regulated in the $Hsp70$ overexpressing clone	15
total genes regulated in Ge-tra (control)	**43**
genes up-regulated in control clone	36
genes down-regulated in control clone	7

that out of the 13 selected genes 9 were regulated only in Ge-tet-1, 2 genes were only regulated in Ge-tra, 1 gene was up-regulated in both clones and one was regulated in both clones but down-regulated in Ge-tet-1 and up-regulated in Ge-tra cells.

5.1 Role of heat shock protein 70 in apopotosis

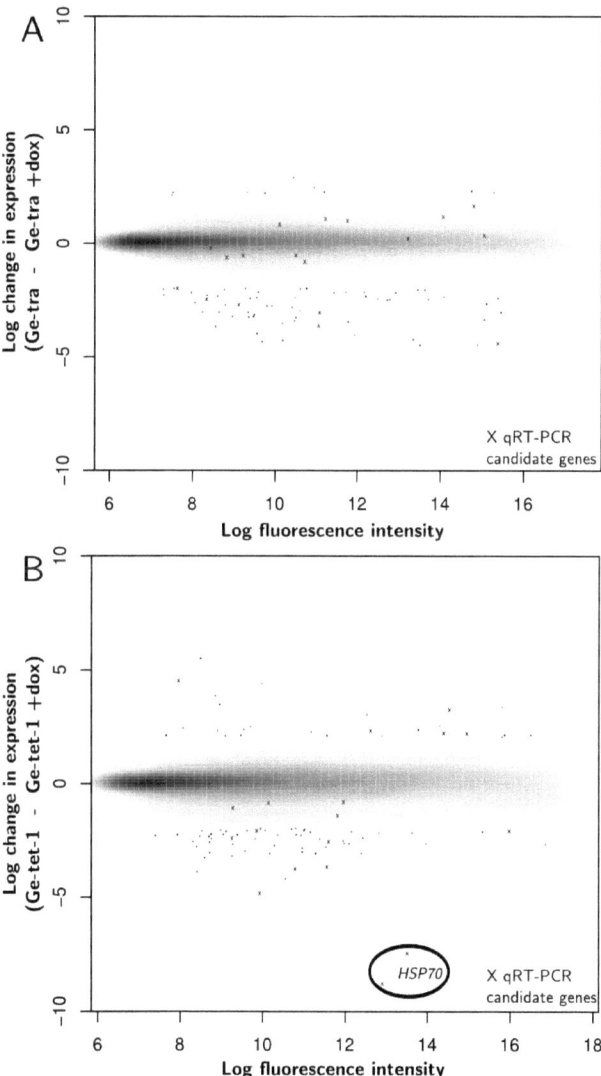

Figure 5.4: Changes in gene expression in Ge-tra and Ge-tet-1 cells upon doxycycline exposure Shown is the expression of 44000 transcripts of genes, with the transcripts of genes with unchanged expression being represented by the grey cloud. The change in expression is given on the y-axis and the fluorescence intensity on the x-axis. Candidate genes, which were analysed in qRT-PCR are indicated with an X. Dots with negative values on the y-axis are upregulated genes upon doxycycline addition and dots with positive values on the y-axis are down-regulated genes. **A** The changes in gene expression in Ge-tra cells after addition of doxycycline for 24 hrs. **B** The change in expression of the *HSP70* gene in Ge-tet-1 cells in comparison to Ge-tet-1 cells treated with doxycycline for 24 hrs is highlighted with a circle. The two X for *HSP70* represent the two 60-mer oligos for *HSP70*, present on the microarray.

5 Results

Table 5.2: Selected genes for validation of microarray data by quantitative real-time PCR Regulated genes with log 2-fold change in expression of smaller than -2 or larger than 2, are listed with gene abbreviation (Abbrev.), gene name (Name), log 2-fold change in expression in Ge-tet versus Ge-tet plus doxycycline (Expr. tet) and Ge-tra versus Ge-tra plus doxycycline (Expr. tra), and the respective p-values (p-val.). Negative values in the change in expression mean an upregulation of the gene upon doxycycline treatment and positive values a downregulation. If for one transcript more than one oligo was present on the microarray all values of expression and p-values were given seperately for the single oligos. The fold change in expression can be calculated from the log 2-fold change in expression by using formula 4.10 on page 65. p-values are significant with a type I error rate α of 0.01. Significant expression and p-values are printed in bold letters.

Abbrev.	Name	Expr. tet	p-val.	Expr. tra	p-val.
ASNS	asparagine synthetase	**-2.12**	**5.57E-04**	**-4.45**	**3.74E-05**
		-2.14	**8.78E-04**	**-4.51**	**5.54E-05**
BCL10	B-cell lymphoma 10	**-2.12**	**5.55E-04**	-0.58	1.00E+00
CASP8	caspase 8, apoptosis-related cysteine peptidase	**4.48**	**3.06E-05**	**-2.02**	**2.05E-04**
EIF5	eukaryotic translation initiation factor 5	**-3.71**	**1.47E-05**	-0.57	1.00E+00
GRB10	growth factor receptor-bound protein 10	-1.44	1.57E-04	**-3.09**	**1.46E-05**
		-0.84	1.00E+00	**-3.66**	**2.64E-05**
HSP70	heat shock protein 1A 70 kDa	**-8.81**	**4.00E-07**	1.03	8.01E-05
		-7.49	**1.10E-06**	0.96	1.00E+00
JUND	jun D proto-oncogene	**-2.58**	**4.61E-04**	-0.84	1.00E+00
SAPS3	SAPS domain family, member 3	**-3.79**	**6.60E-04**	0.79	1.00E+00
STC2	stanniocalcin 2	-0.90	1.00E+00	**-2.76**	**3.55E-05**
		-1.12	2.39E-03	**-2.51**	**1.06E-04**
TBL1XR1	transducin (beta)-like 1X-linked receptor 1	**-2.46**	**8.04E-05**	0.43	1.00E+00
		-2.43	**1.59E-03**	-0.24	1.00E+00
THOC4	THO complex 4	**2.28**	**1.77E-03**	0.18	1.00E+00
		3.20	**2.36E-03**	0.31	1.00E+00
TM4SF1	transmembrane 4 L six family member 1	**2.17**	**1.68E-04**	1.14	**1.14E-03**
TXNRD1	thioredoxin reductase 1	**-4.87**	**4.50E-04**	-0.66	1.00E+00

A list of all 118 regulated genes can be found in the appendix (table B.2 on page 176).

5.1 Role of heat shock protein 70 in apoptosis

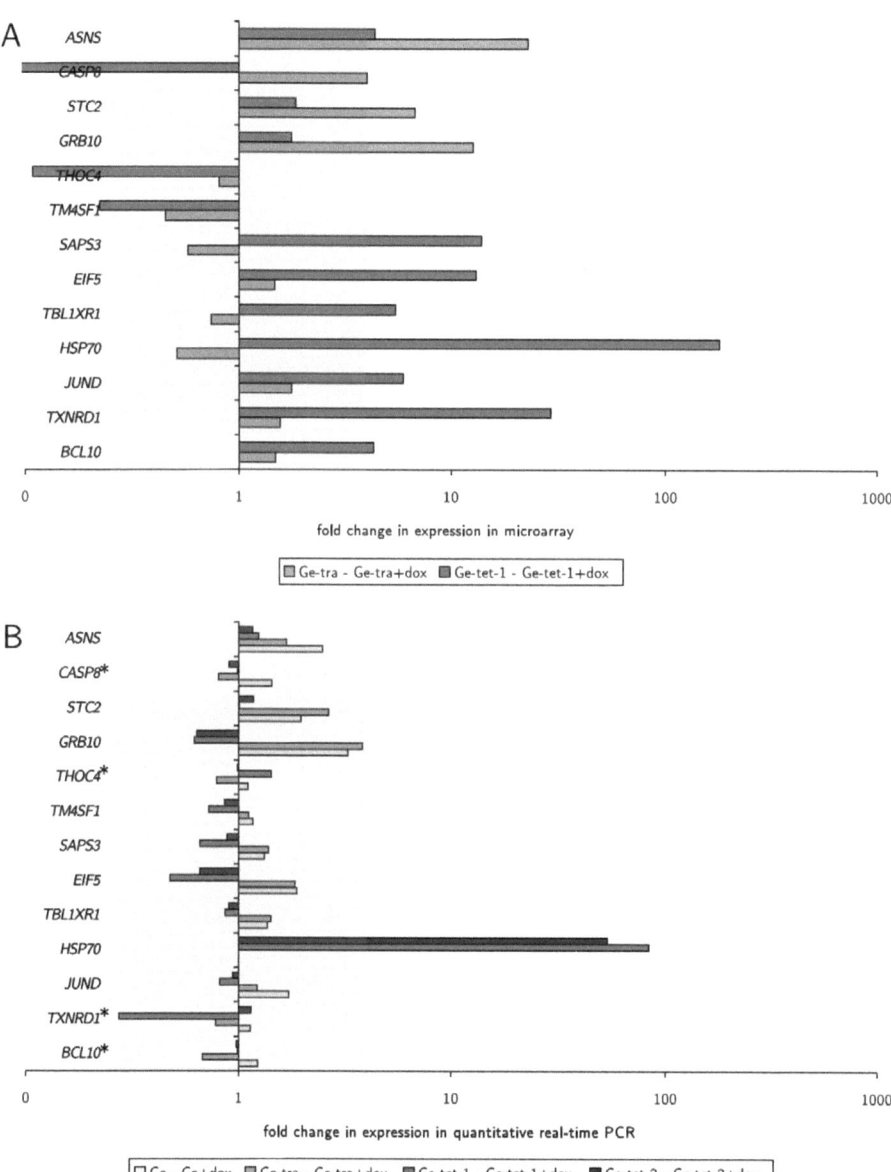

Figure 5.5: Changes in expression of 13 selected genes as determined by a whole human microarray analysis and by qRT-PCR The light grey box represents the applied threshold of 0.25 to 4.00-fold change in expression, which was used in the microarray analysis. Bars reaching over the box are regulated above this threshold. **A** The results of the microarray analysis **B** Each bar represents the result of the mean value of a triplicate of ct values calculated with formula 4.11 on page 65. Genes marked with an asterisk (*) showed contrary tendencies of expression in the control clones Ge and Ge-tra or in the inducible clones Ge-tet-1 and 2, or both within the qRT-PCR analysis. Thus, their regulation appeared to be inconsistant.

5 Results

5.1.2.2 Quantitative real-time PCR analysis of selected genes

The same isolated RNA used in the microarray analysis was also used for the qRT-PCR and additionally RNA was isolated from the parental cell line Ge and another inducible clone, Ge-tet-2. For the expression of HSP70 in the different clones used for RNA isolation see figure 5.3 on page 70. The qRT-PCR was performed as described in section 4.4.3 on page 60 in the methods part. The change in expression was calculated with formula 4.11 on page 65 and is presented for the selected genes in part B of figure 5.5 on the previous page.

One quality control of the results of qRT-PCR analysis are the dissociation curves as seen in figure A.1 on page 173. If the curves of the cDNA of one clone amplified with the same primer overlap, this is an indicator that these amplicons all have the same melting behaviour and are identical, so that mispriming or other artefacts can be excluded. It is furthermore essential for the quality of the data, that nothing is amplified in the negative controls, where only the SYBR green mix and the primers were added, without template cDNA. In all reactions performed, the negative controls were indeed negative.

The analysis of the 13 selected genes by qRT-PCR revealed that in 9 out of 13 genes, the change in expression showed the same tendency (up- or down-regulation) within the control clones Ge and Ge-tra or within the two inducible clones Ge-tet-1 and Ge-tet-2. The genes in which the expression tendency in both control clones or both inducible clones compared with each other within the qRT-PCR results was not identical, were marked with an asterisk (*) in part B of figure 5.5 on the previous page. Nevertheless, by comparing the data of the qRT-PCR with the ones of the microarray analysis the following results were gained: for the expression in Ge-tet-1 cells in the qRT-PCR the tendency of expression, meaning up- or down-regulation of genes, was the same as in the microarray in 5 out of 13 genes. For Ge-tra the same expression tendency for qRT-PCR and microarray was found in 6 out of 13 genes. No qRT-PCR data are shown for *Hsp70* in the Ge and the Ge-tra clone, as the ct-values were above 30 and were therefore neglected, as a ct above 30 is considered to indicate unspecific amplification. Thus, except for the expression of the internal control *Hsp70*, induced by the addition of doxycycline in Ge-tet-1 and Ge-tet-2, no other gene was regulated in such a way that its fold change in expression in qRT-PCR was below or above the threshold of 0.25 to 4.00-fold change in expression applied to the microarray data.

In summary, only one change in expression of the microarray analysis out of 13 investigated genes could be clearly confirmed by the qRT-PCR, which was the acute overexpression of *Hsp70* itself. The acute overexpression of *Hsp70* therefore likely does not improve CTL-induced apoptosis by induction of pro-apoptotic genes on the transcriptome level. This effect of HSP70 is more likely mediated on the protein level. Thus, key steps in apoptosis were further analysed on the protein level in cells acutely overexpressing HSP70 and control clones.

To further elucidate the molecular mechanisms behind the increased susceptibility of tumour

5.1 Role of heat shock protein 70 in apopotosis

cells acutely overexpressing HSP70 towards CTLs, GrB was used as a molecular trigger of apoptosis.

5.1.3 Effect of acute and permanent overexpression of heat shock protein 70 on early and late apoptosis

5.1.3.1 Effect of acute HSP70 overexpression on phosphatidylserine exposure after granzyme B-induced apoptosis

To determine, whether GrB, a component of the cytotoxic granules of CTLs, would differentially trigger apoptosis in Ge-tet cells acutely overexpressing HSP70 in comparison to allogeneic CTLs, the following experiment was performed. An early hallmark of apoptosis is the externalisation of PSs to the cell surface, which are normally located at the inner layer of the cellular membrane. The presence of PS on the cell surface allows detection by annexin V, which binds to PS in a calcium-dependent manner. The co-staining with PI allows to distinguish between early apoptotic and late apoptotic or necrotic cells. For the flow cytometric analysis the percentage of annexin V/PI (upper right quadrant) and annexin V-positive cells (lower right quadrant) as illustrated in figure 5.6 is summarised.

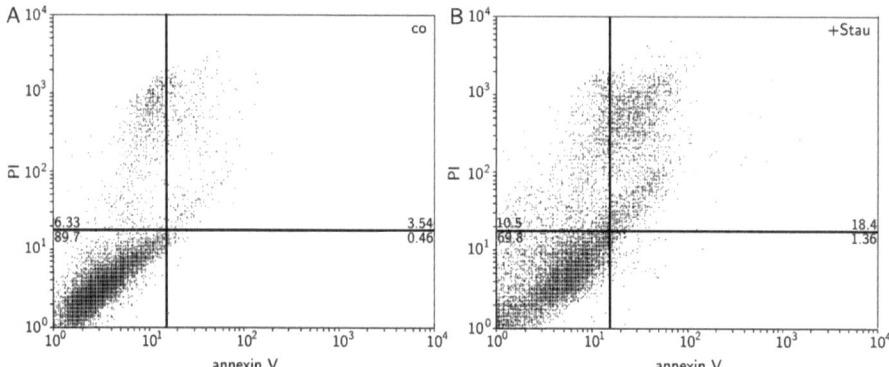

Figure 5.6: **Flow cytometric analysis of annexin V/PI staining** Apoptosis was induced by the addition of 1 μM staurosporine (+Stau) for 20 hrs. Control cells (co) were left untreated. Cells were stained with annexin V/PI for 20–30 min at 4 °C and were subsequently analysed by flow cytometry. For the evaluation of the data, the percentage of annexin V (lower right quadrant) and annexin V/PI (upper right quadrant)-positive cells was evaluated. The percentage of cells represented in all quadrants is given in the dot plots.

It was investigated, whether the acute overexpression of HSP70 had an influence on annexin V staining in GrB-induced apoptosis. Doxycycline was added to the cells for 24 hrs and the acute overexpression of HSP70 in Ge-tet clones was determined by intracellular flow cytometry

5 Results

afterwards (part A of figure 5.7). After inducing apoptosis by GrB for 24 hrs, cells were analysed for annexin V/PI staining (part B of figure 5.7).

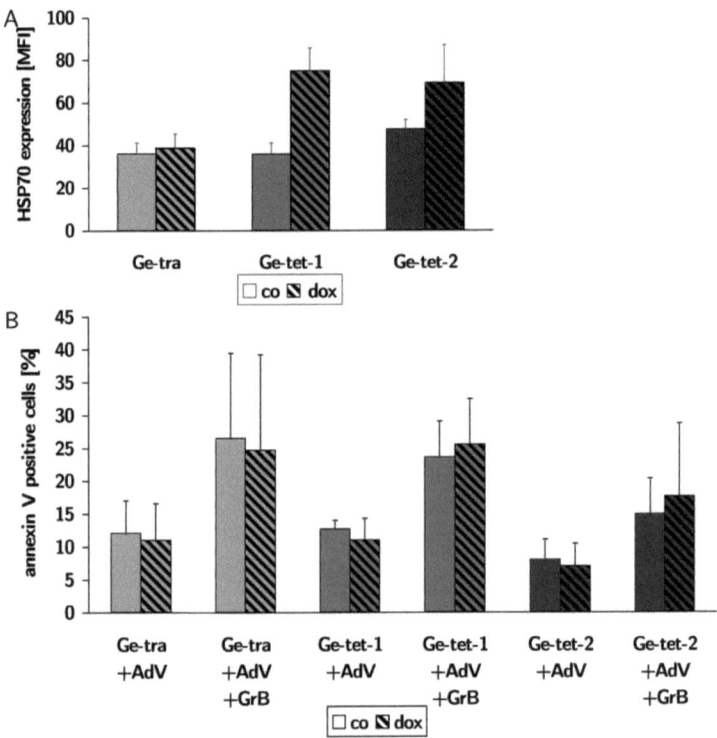

Figure 5.7: Acute HSP70 overexpression did not influence GrB-induced apoptosis as measured by binding of annexin to exposed phosphatidylserines Ge clones were first treated with doxycycline and then apoptosis was induced using GrB and AdV. **A** HSP70 expression levels after addition of doxycycline (dox) for 24 hrs were determined by intracellular flow cytometry. Control cells (co) were left untreated. The mean values of 4 independent experiments are shown, error bars depict standard error of the mean (SEM). **B** After confirmation of HSP70 induction in the Ge-tet clones, AdV and 1 ng/μl GrB (+AdV +GrB) or AdV only as control (+AdV) were added to the cells for 24 hrs. Cells were harvested with PBS/EDTA and stained with annexin V-FITC and PI to determine the percentage of apoptotic cells. Striped bars represent cells treated with doxycycline (dox), control (co) cells were left untreated. Mean values of 4 independent experiments are shown, error bars depict SEM.

The exposure of PS was not changed in cells acutely overexpressing HSP70 in comparison to control cells after GrB-induced apoptosis as determined by annexin V/PI staining. It must be noted, that the induction of HSP70 in the Ge-tet-2 clone was relatively low in comparison to the Ge-tet-1 clone in these experiments. Importantly, HSP70 did not protect Ge-tet-1 and Ge-tet-2 cells from GrB-induced apoptosis.

5.1.3.2 Effect of acute HSP70 overexpression on granzyme B-induced DNA fragmentation

To investigate, whether instead of the early apoptotic hallmark, the externalisation of PS to the cell surface, a late apoptotic hallmark, the fragmentation of DNA, was influenced by the acute overexpression of HSP70, the following experiments were performed. DNA fragmentation and condensation can be analysed by sub G1-peak analysis in flow cytometry. In this assay the percentage of cells with a fluorescence below the signal of the cells in G1-phase is analysed (figure 5.8).

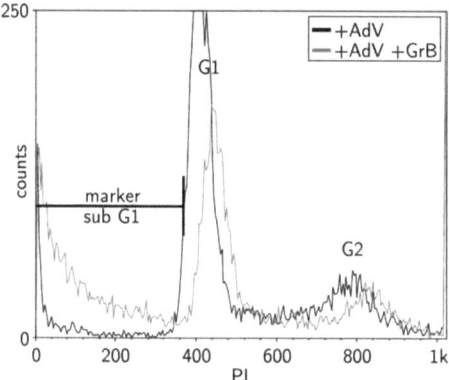

Figure 5.8: Flow cytometric analysis of the sub G1 peak Apoptosis was induced by the addition of AdV and 1 ng/μl GrB (+AdV +GrB), or as control, just the AdV (+AdV) was added for 24 hrs. Cells were fixed and stained with a PI. RNase A was added to degrade RNA and subsequently cells were analysed by sub G1-peak analysis in flow cytometry. For the evaluation of the data, the percentage of cells with a fluorescence below the one of the G1-peak (under the marker) were evaluated. Indicated is also the G2-peak.

HSP70 was acutely overexpressed by the addition of doxycycline for 24 hrs in Ge-tet cells. In the control clone, Ge-tra, the addition of doxycycline had no effect on HSP70 expression (part A of figure 5.9 on the following page). After confirmation of acute overexpression of HSP70 in Ge-tet clones by flow cytometry, GrB and AdV, or as control AdV only, were added to the Ge cells for 24 hrs. The percentage of apoptotic cells with a lower DNA content due to DNA fragmentation and chromatin condensation was determined by sub G1-peak analysis in flow cytometry (part B of figure 5.9 on the next page).

In GrB-induced apoptosis, the acute overexpression of HSP70 in target cells had again no protective effect. Interestingly, HSP70 seemed to further improve apoptosis as determined by DNA fragmentation measured by sub G1-peak analysis. Ge-tet-2 cells showed an increase in the percentage of apoptotic cells upon acute overexpression of HSP70 ($p = 0.01$). However, the percentage of apoptotic cells did not increase significantly in the Ge-tet-1 clone ($p = 0.22$). The

5 Results

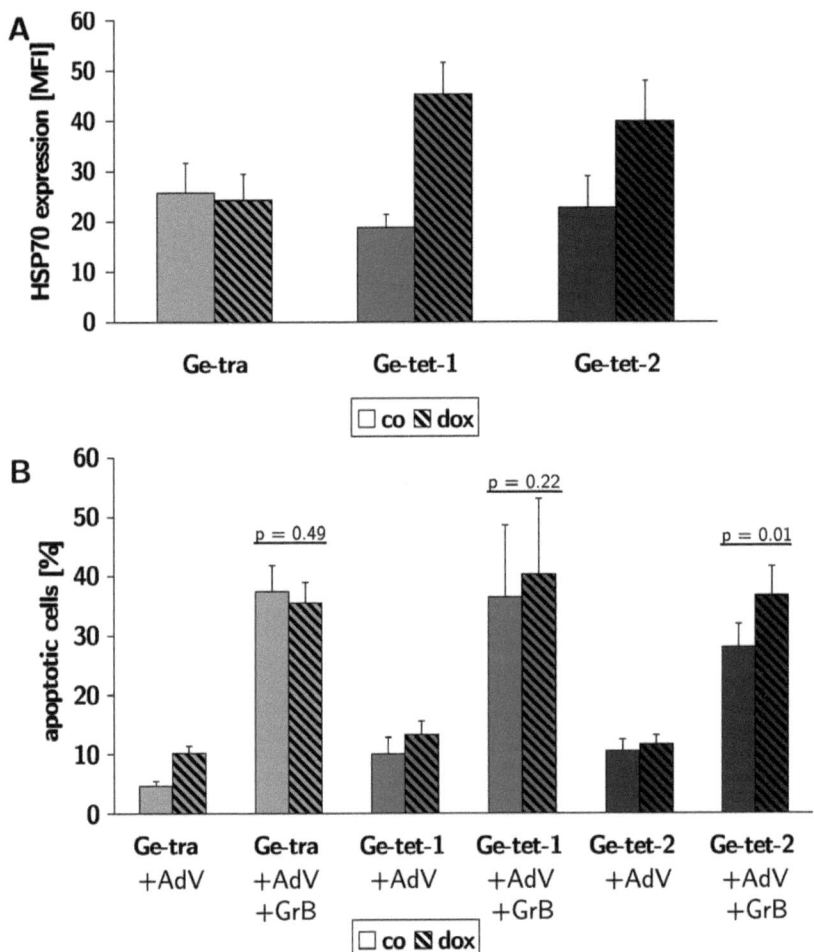

Figure 5.9: Acute HSP70 overexpression can enhance GrB-induced apoptosis as determined by sub G1-peak analysis Ge clones were first treated with doxycycline and then apoptosis was induced using GrB and AdV. **A** Treatment of Ge-tra and Ge-tet cells with 5 µg/ml doxycycline (dox; striped bars) for 24 hrs induces acute HSP70 overexpression in both Ge-tet clones. As control (co) no doxycycline was added. Shown are the mean values of 7 independent doxycycline treatments. These cells were used for the induction of apoptosis with GrB. Error bars depict the SEM. **B** After confirmation of HSP70 induction in the Ge-tet clones either AdV-GFP alone at an MOI of 500 as control, or AdV-GFP and 1 ng/µl human GrB were added. After 24 hrs at 37 °C cells were harvested for subsequent sub G1-peak measurement in flow cytometry. Striped bars represent cells treated with doxycycline (dox), control (co) cells were left untreated. Mean values of 7 independent experiments are shown. Error bars depict SEM. An asterisk (*) indicates a significant difference.

5.1 Role of heat shock protein 70 in apopotosis

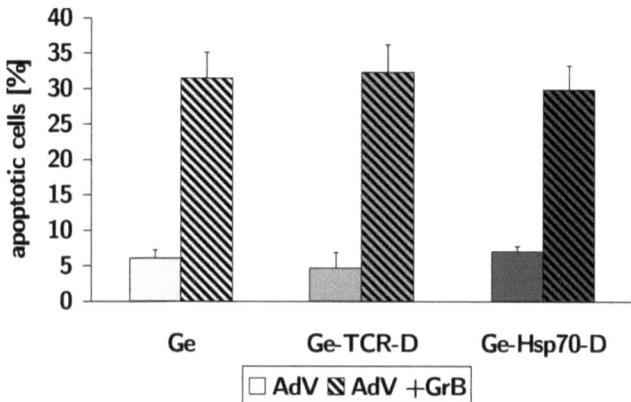

Figure 5.10: Permanent overexpression of HSP70 does not influence GrB-induced apoptosis as determined by sub G1-peak analysis Cells were seeded in 24-well plates. Either AdV-GFP was added alone with a MOI of 500 or together with 1 ng/µl GrB (striped bars) for 24 hrs at 37 °C. Medium was collected, cells were harvested using trypsin/PBS/EDTA, fixed in ice-cold EtOH, and stained with PBS/RNase A/PI for 30 min at 37 °C for subsequent sub G1-peak analysis in flow cytometry. Depicted are the mean values of 3 independent experiments with error bars representing SEM.

control clone Ge-tra did not show an increase of apoptotic cells after addition of doxycycline (p = 0.49). The p-values were determined with the Wilcoxon-test. Thus, acute overexpression of HSP70 appears in principle to be able to increase the progression of GrB-induced apoptosis to DNA fragmentation although not in all clones analysed.

5.1.3.3 Effect of permanent HSP70 overexpression on granzyme B-induced DNA fragmentation

To investigate, whether late stages of apoptosis induced by GrB are also affected by permanent overexpression of HSP70, the following experiment was performed. For the induction of apoptosis by GrB Ge cells permanently overexpressing HSP70 (Ge-Hsp70-D) were used. As control, the parental cell line Ge without any construct or Ge-TCR-D cells, permanently overexpressing the β-chain of a rat TCR, were used. GrB together with AdV-GFP, or as control, AdV-GFP alone, were added to the cells. After 24 hrs a sub G1-peak measurement was performed by flow cytometry. The results are illustrated in figure 5.10.

The permanent overexpression of HSP70, in contrast to the acute overexpression, did not influence the susceptibility of tumour cells towards apoptosis induced by GrB. Noteworthy, also permanent overexpression of HSP70 did not confer protection against GrB. Statistical analysis of a comparison of all three clones using the H-test (Kruskal-Wallis) did not show a significant difference between the clones (p = 0.9).

5 Results

5.1.3.4 Effect of acute HSP70 overexpression on granzyme B uptake

Next, it was investigated whether the significant increase in late apoptotic cells found in the Ge-tet-2 clone acutely overexpressing HSP70, was due to an enhanced uptake of GrB into these cells. It was shown earlier that GrB and HSP70 are able to interact and it was also reported that cell surface-bound HSP70 is able to mediate a perforin-independent uptake of GrB (Gross et al. 2003b). Furthermore, permanently overexpressed HSP70 was shown to bind to clathrin in Ge cells (Dressel et al. 2003) suggesting a role for HSP70 in clathrin-dependent endocytosis. To investigate whether acute HSP70 overexpression influences the uptake of GrB, the expression of HSP70 was induced for 24 hours with doxycycline in both Ge-tet clones (Ge-tet-1 and Ge-tet-2). In the control cell line (Ge-tra), transfected just with the transactivator system, the addition of doxycyline did not induce HSP70 overexpression.

The induction of HSP70 by doxycycline led to an increase in HSP70 compared to normal levels in the Ge-tet clones (part A and B of figure 5.11 on the next page). Alexa-488-labelled GrB was added afterwards to the cells. The uptake of labelled GrB was determined by flow cytometry (part C and D of figure 5.11 on the facing page).

The flow cytometric analysis showed no change in GrB uptake after acute overexpression of HSP70, although there was an increase in HSP70 expression (parts A and B of figure 5.11 on the next page). Thus, acutely overexpressed HSP70 did apparently not improve CTL or GrB-mediated apoptosis by increasing the uptake of GrB.

5.1.3.5 Effect of acute HSP70 overexpression on phosphatidylserine exposure after staurosporine-induced apoptosis

To determine whether the improvement of GrB-induced apoptosis in Ge-tet-2 cells by the acute overexpression of HSP70 was specific for GrB or whether an increased percentage of apoptotic cells would also be found upon other apoptotic stimuli, staurosporine was tested as an inducer of early and late apoptosis. Ge-tra and Ge-tet cells were treated with doxycycline for 24 hrs. After confirmation of acute HSP70 overexpression in Ge-tet cells by flow cytometry (part A of figure 5.12 on page 82), medium containing 1 μM staurosporine and doxycycline were added to the cells for 20 hrs. The percentage of apoptotic cells was determined by annexin V/PI staining in flow cytometry (part B of figure 5.12 on page 82).

The acute overexpression of HSP70 seemed to be able to partly protect cells from staurosporine-induced apoptosis. The percentage of early apoptotic cells was significantly reduced in Ge-tet-1 cells ($p = 0.04$), although not in Ge-tet-2 cells ($p = 0.46$). The amount of apoptotic cells was also not significantly changed in the control clone Ge-tra ($p = 0.17$). The statistical analysis was made with the Wilcoxon-test.

5.1 Role of heat shock protein 70 in apopotosis

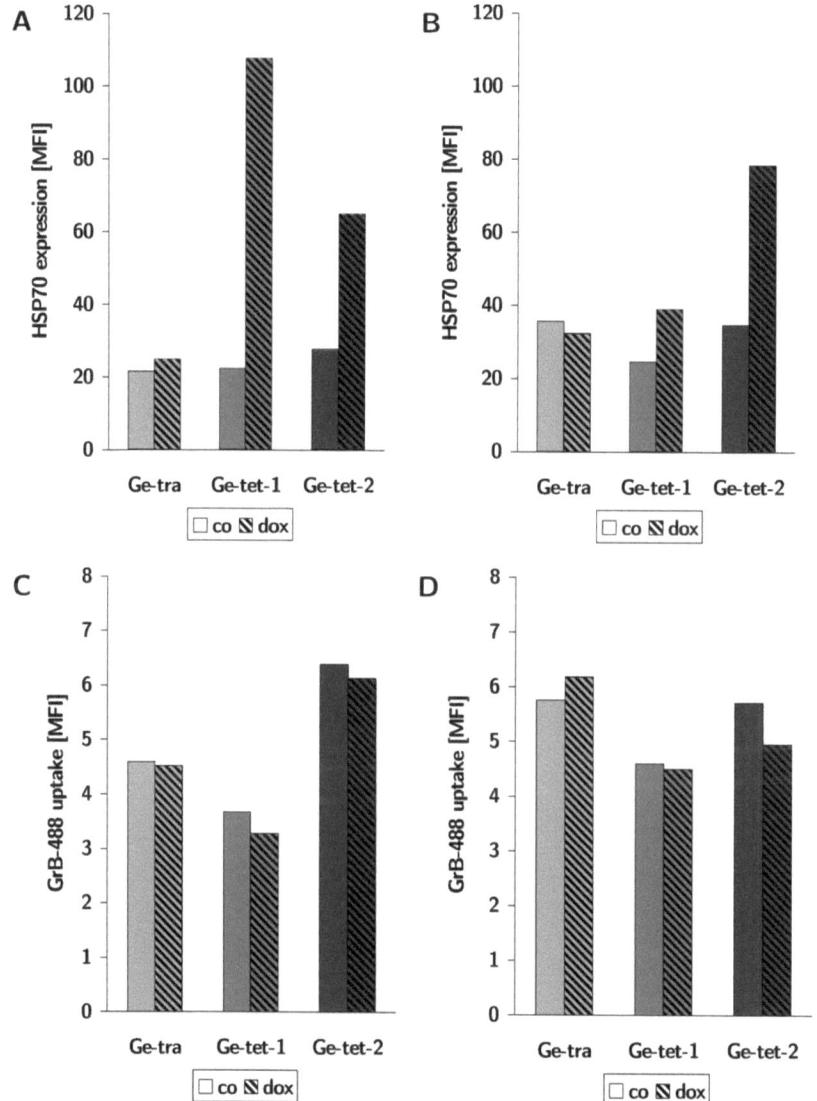

Figure 5.11: Acute HSP70 overexpression does not influence GrB-uptake Ge clones were treated with doxycycline for 24 hrs. Then Alexa-488 labelled GrB was added and subsequently cells were analysed by flow cytometry. Two independent experiments are shown in this figure. **A** and **B** Treatment of Ge-tra and Ge-tet cells with 5 µg/ml doxycycline (dox; striped bars) for 24 hrs induces acute HSP70 overexpression in both Ge-tet clones. As control (co) no doxycycline was added. **C** and **D** 1 ng/µl human GrB was added to the cells for 1 hr at 37 °C. Afterwards, cells were washed twice with trypsin/EDTA/PBS to remove unbound GrB and were subsequently analysed by flow cytometry.

5 Results

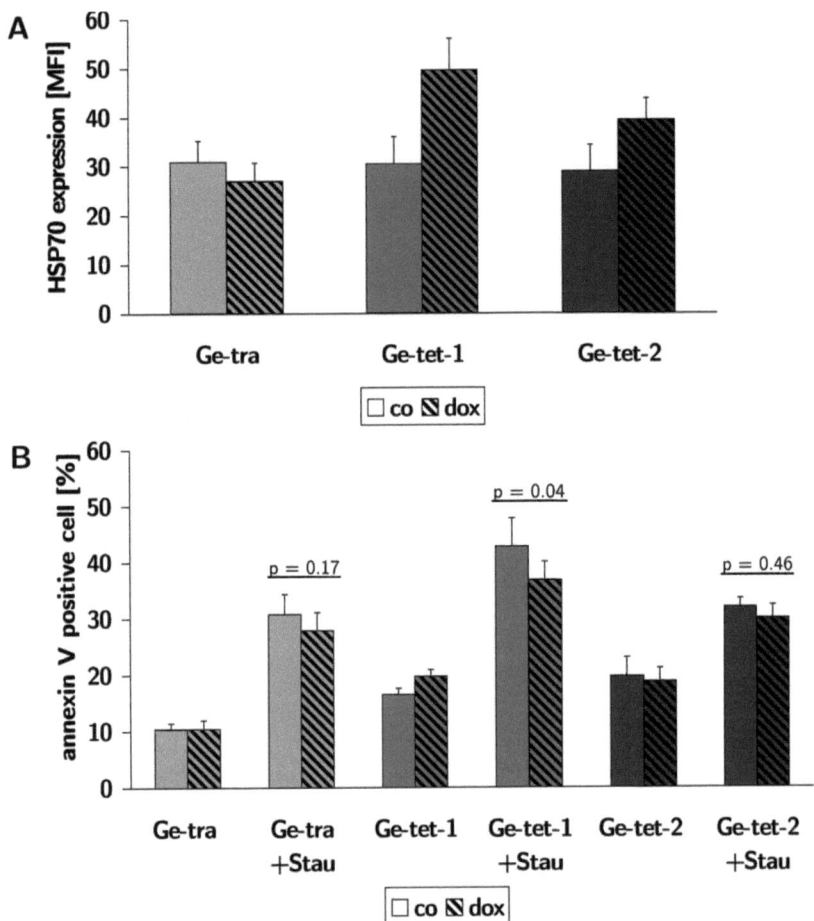

Figure 5.12: Acute HSP70 overexpression partially protects from staurosporine-induced apoptosis Ge clones were first treated with doxycycline and then apoptosis was induced using staurosporine. **A** Treatment of Ge-tra and Ge-tet cells with 5 µg/ml doxycycline (dox; striped bars) for 24 hrs induced acute HSP70 overexpression in both Ge-tet clones. Control (co) cells were left untreated. Shown are the mean values of 12 independent doxycycline treatments with Ge-tra, 8 treatments with Ge-tet-1 and 6 treatments with Ge-tet-2. The same cells were used for the induction of apoptosis with staurosporine. Error bars depict SEM. **B** After confirmation of HSP70 induction in the Ge-tet clones, 1 µM staurosporine (+Stau) was added for 20 hrs. After induction of apoptosis cells were harvested and stained with annexin V-FITC and PI for apoptotic cells. Striped bars represent cells treated with doxycycline (dox), control (co) cells were left untreated. The mean values of 12 independent experiments with Ge-tra, 8 experiments with Ge-tet-1 and 6 experiments with Ge-tet-2 are shown. Error bars depict SEM. Given are the p-values determined with a Wilcoxon-test. Significant p-values are indicated with an asterisk (*).

5.1.3.6 Effect of acute HSP70 overexpression on staurosporine-induced DNA fragmentation

To investigate, whether the acute overexpression of HSP70 would also have an effect on late apoptosis meaning DNA fragmentation induced by staurosporine, a sub G1-peak analysis was performed. Cells were treated with doxycycline for 24 hrs and acute overexpression of HSP70 was analysed in both Ge-tet clones (part A of figure 5.13 on the following page). Apoptosis was induced by the addition of staurosporine for 20 hrs and DNA fragmentation was analysed by sub G1-peak analysis in flow cytometry (part B of figure 5.13 on the next page).

In contrast to the effect of acute overexpression of HSP70 in GrB-induced apoptosis, acute HSP70 overexpression did not increase the percentage of apoptotic cells in staurosporine-induced apoptosis. However, in this experimental series the induction of HSP70 was rather poor in the Ge-tet-2 clone.

5.1.3.7 Effect of permanent HSP70 overexpression on staurosporine-induced DNA fragmentation

Next, it was tested whether the partially protective effect against staurosporine-induced apoptosis in the Ge-tet-1 clone acutely overexpressing HSP70 would also be found in cells permanently overexpressing HSP70 by sub G1-peak analysis. For this purpose, Ge cells permanently overexpressing HSP70 or as control the β-chain of a rat TCR, and the parental cell line Ge were seeded and 1 μM staurosporine was added for 20 hrs. The percentage of apoptotic cells was determined by sub G1-peak analysis (figure 5.14 on page 85). The permanent overexpression of HSP70 did not alter DNA fragmentation in staurosporine-induced apoptosis. Statistical analysis using the Kruskal-Wallis-test indicated that there is no significant difference between the three Ge-clones in DNA fragmentation after staurosporine induced apoptosis ($p = 0.9$). No effect of permanent HSP70 overexpression was found on PS externalisation after staurosporine-induced apoptosis (data not shown).

To investigate, which key steps in apoptosis are involved in the increase in DNA fragmentation seen in Ge-tet-2 cells after acute overexpression of HSP70 during GrB-induced apoptosis, further experiments were performed.

5.1.4 Influence of acute HSP70 overexpression on key steps in apoptosis

To analyse which key steps in apoptosis (figure 5.15 on page 86) were influenced by the acute overexpression of HSP70 in GrB-induced apoptosis on the protein level, additional experiments were performed. Apoptosis can be induced either by the intrinsic pathway involving mitochondria or by the classical extrinsic pathway involving caspases. Both pathways are linked at several points with each other. Two of the key steps of the intrinsic pathway of apoptosis

5 Results

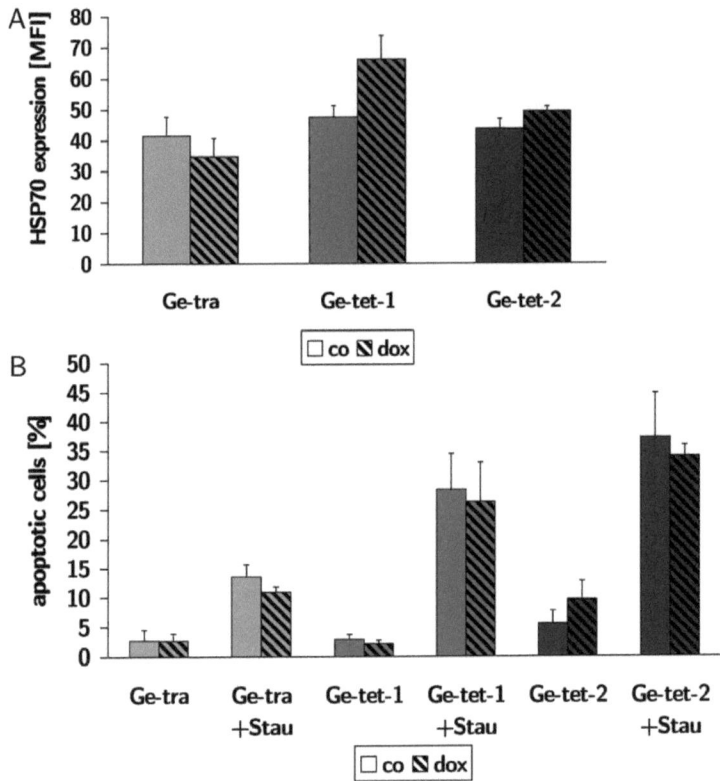

Figure 5.13: Acute HSP70 overexpression does not improve staurosporine-induced apoptosis as determined by sub G1-peak analysis Ge clones were first treated with doxycycline, then apoptosis was induced by staurosporine. **A** HSP70 was induced by 5 µg/ml doxycycline (dox) in Ge-tet clones for 24 hrs or cells were left untreated (co). The induction of HSP70 was analysed by intracellular staining for flow cytometry. Shown are the results of 3 independent experiments with error bars depicting SEM. **B** Apoptosis was induced by the addition of 1 µM staurosporine (+Stau) for 20 hrs. Medium was collected, cells were harvested and fixed in ice-cold EtOH before they were stained for flow cytometric analysis with PBS/RNase A/PI for 30 min at 37 °C for sub G1-peak analysis. The results of 3 independent experiments are shown, error bars depict SEM.

are the loss of mitochondrial membrane potential $\Delta\Psi$ and the release of cytochrome c from mitochondria. Key steps of the extrinsic apoptotic pathway are the activation of the initiator caspase-8 and the activation of the effector caspase-3, which can be activated by proteolysis by molecules of both the intrinsic and extrinsic pathway. In the nucleus caspase-3 can cleave ICAD, which then releases CAD, and the latter induces DNA fragmentation (Neimanis et al. 2007).

5.1 Role of heat shock protein 70 in apopotosis

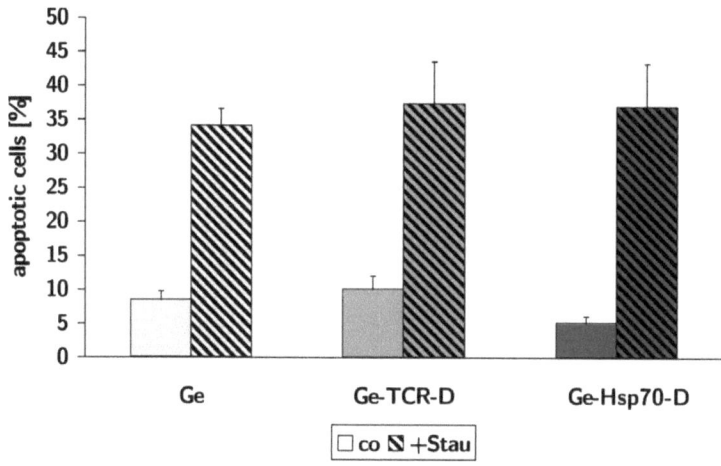

Figure 5.14: Permanent overexpression of HSP70 does not influence staurosporine-induced apoptosis as determined by sub G1-peak analysis For the induction of apoptosis 1 μM staurosporine (+Stau) was added to the cells. Control cells (co) were left untreated. After 20 hrs, cells were harvested for subsequent sub G1-peak analysis in flow cytometry. Mean values of 4 independent experiments are shown, error bars depict SEM.

5.1.4.1 Effect of acute HSP70 overexpression on the change in mitochondrial membrane potential $\Delta\Psi$

One of the hallmarks of apoptosis in the intrinsic pathway is the loss of the mitochondrial membrane potential $\Delta\Psi$. This change in ion distribution over the mitochondrial membrane can be visualised by JC-1, a dye, which forms aggregates, when there is a high $\Delta\Psi$ and is present as monomer upon a loss of $\Delta\Psi$. The fluorescence of those cells shifts from 590 nm (FL-2) to 527 nm (FL-1) in flow cytometry (figure 5.16 on page 87).

To test whether the acute overexpression of HSP70 had an effect on the change in $\Delta\Psi$ in GrB-induced apoptosis, the following experiments were performed. Ge-tra and Ge-tet cells were treated with doxycycline for 24 hrs to induce HSP70 expression in Ge-tet clones (part A of figure 5.17 on page 88). GrB and AdV-β-gal, or as control only AdV-β-gal, were added to the cells for 24 hrs for the induction of apoptosis. Cells were stained with JC-1, washed 2 times with ice-cold PBS and subsequently fixed with PFA. The percentage of viable (high $\Delta\Psi$) and apoptotic cells (low $\Delta\Psi$) was determined by flow cytometry. A summary of the percentage of cells with reduced $\Delta\Psi$ of 4 experiments is given in part B of figure 5.17 on page 88.

The acute overexpression of HSP70 did not change the percentage of cells, which had a reduced $\Delta\Psi$ upon delivery of GrB.

Next, it was analysed, whether the acute overexpression of HSP70 would affect the $\Delta\Psi$ in staurosporine-induced apoptosis. Cells were again treated with doxycycline for 24 hrs and

5 Results

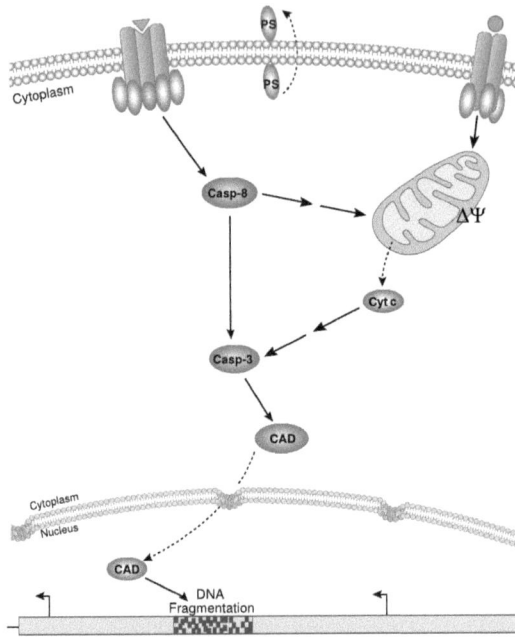

Figure 5.15: Analysed key steps in apoptosis The externalisation of PS from the inner leaflet of the cell membrane to the cell surface was already investigated in the last chapter, as well as the DNA fragmentation by sub G1-peak analysis in flow cytometry. The loss of mitochondrial membrane potential ($\Delta\Psi$) and the relase of cytochrome c (Cyt c) from mitochondria were further analysed in flow cytometry. The activation of caspase-8 (Casp-8) and the activation of caspase-3 (Casp-3) were further investigated by immunoblot and flow cytometry, respectively. DNA fragmentation induced by CAD was further assessed by the presence of the apoptotic ladder after electrophoretic separation of DNA on agarose gels. The figure was adapted from (Cellsignal 2008).

HSP70 was induced in both Ge-tet-clones (part A of figure 5.18 on page 89). Apoptosis was induced using staurosporine for 20 hrs and $\Delta\Psi$ was measured in flow cytometry after staining with JC-1 (part B of figure 5.18 on page 89).

Also in staurosporine-induced apoptosis, the acute overexpression of HSP70 did not change the amount of cells with a loss in $\Delta\Psi$. Thus, the acute overexpression of HSP70 did not seem to have an influence on GrB and staurosporine-induced change of $\Delta\Psi$.

5.1.4.2 Effect of acute HSP70 overexpression on release of cytochrome c from mitochondria

The second hallmark of the intrinsic apoptotic pathway, which was investigated, was the release of cytochrome c from mitochondria. Cytochrome c is a heme protein belonging to the electron

5.1 Role of heat shock protein 70 in apopotosis

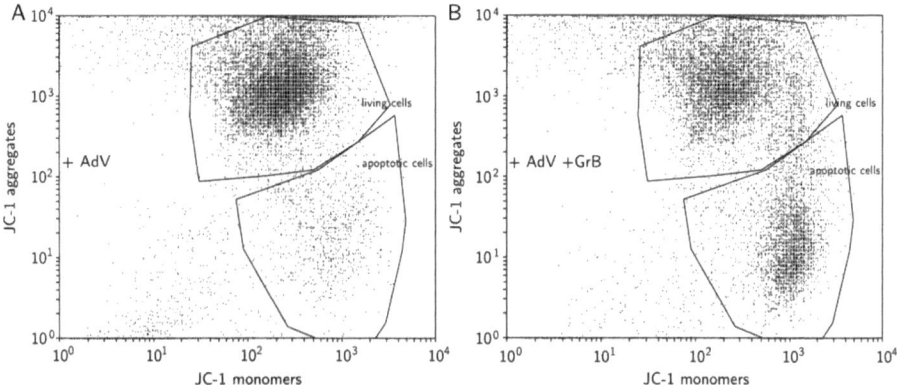

Figure 5.16: Determining the mitochondrial membrane potential $\Delta\Psi$ by flow cytometry GrB-induced apoptosis in Ge-tet-1 cells increased the number of cells with reduced $\Delta\Psi$ as determined in flow cytometry by JC-1 staining. Shown are two dot plots with Ge-tet-1 cells stained with JC-1. **A** As a control AdV-β-gal (+AdV) was added to the cells for 24 hrs. **B** For induction of apoptosis GrB and AdV-β-gal (+AdV +GrB) were added for 24 hrs.

transport chain and is loosely associated with the inner mitochondrial membrane. In the flow cytometric analysis the release of cytochrome c from mitochondria was evaluated as an increase of cells with a lower content of mitochondrial cytochrome c upon induction of apoptosis by AdV and GrB (Stahnke et al. 2004). This increase in the percentage of cells with a reduced mitochondrial cytochrome c content was determined as shown in figure 5.19 on page 90.

It was tested, whether the acute overexpression of HSP70 had an effect on the release of cytochrome c in GrB-induced apoptosis. Acute HSP70 overexpression was induced in Ge-tet clones after the addition of doxycycline for 24 hrs (part A of figure 5.20 on page 91). Apoptosis was induced by the addition of AdV-β-gal and GrB for 24 hrs. The release of mitochondrial cytochrome c is presented as percentage of cells with reduced mitochondrial cytochrome c content and was determined with a cytochrome c-specific antibody (part B of figure 5.20 on page 91).

There seemed to be a tendency that in GrB-induced apoptosis in the Ge-tet-1 clone the acute overexpression of HSP70 further increased the proportion of cells with reduced mitochondrial cytochrome c. Statistical analysis using the Wilcoxon-test determined a p-value of 0.06 for the Ge-tet-1 clones in GrB induced apoptosis between cells acutely overexpressing HSP70 and cells with normal HSP70 expression. This effect was not seen in the other inducible clone Ge-tet-2 (p = 0.35). Notably, in the Ge-tet-2 clone, the overall amount of cells with reduced mitochondrial cytochrome c content did increase less upon addition of GrB-induced apoptosis compared to both other clones.

Next, it was determined, whether the same or a contrary tendency would be seen upon

5 Results

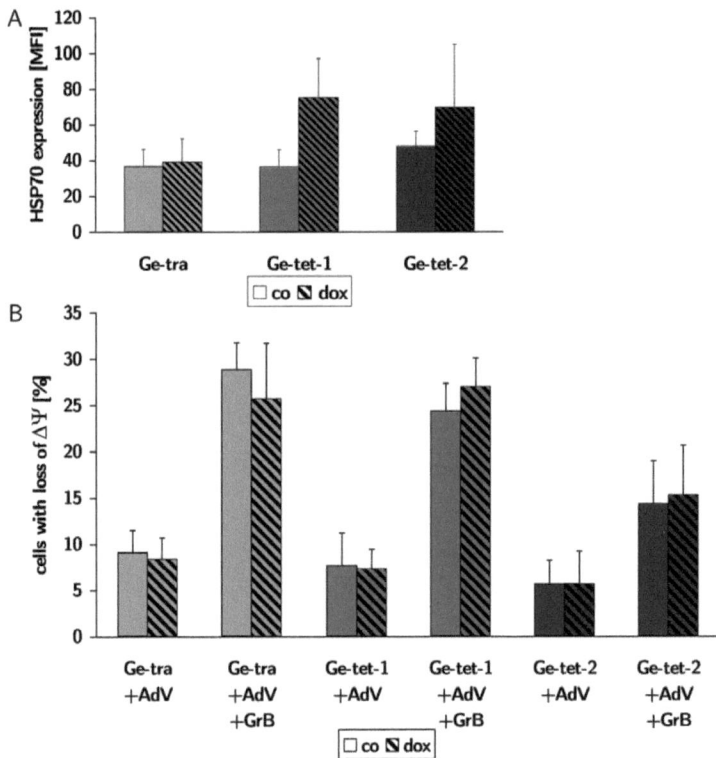

Figure 5.17: Acute HSP70 overexpression did not influence the change in mitochondrial membrane-potential after GrB-induced apoptosis Acute induction of HSP70 in Ge-tet clones and GrB-induced change of $\Delta\Psi$. **A** Confirmation of acute HSP70 overexpression in the Ge-tet clones after addition of doxycycline (dox). Control (co) cells were left untreated **B** 1 ng/μl GrB (+GrB) was added together with AdV-β-gal (+AdV) for 24 hrs for induction of apoptosis. As a control just AdV-β-gal (+AdV) was added. Cells were harvested, stained with JC-1 and fixed for flow cytometric measurement. Shown is the percentage of cells with reduced $\Delta\Psi$. Depicted are the mean values of 4 independent experiments with error bars indicating SD.

staurosporine-induced apoptosis. Doxycycline was again added to Ge-tra and Ge-tet clones for the acute overexpression of HSP70 in the inducible Ge-tet clones (part A of figure 5.21 on page 92). Apoptosis was elicited upon the addition of staurosporine for 20 hrs. The percentage of cells with a reduced amount of mitochondrial cytochrome c was analysed by flow cytometry (part B of figure 5.21 on page 92).

The acute overexpression of HSP70 in the Ge-tet-1 and Ge-tet-2 clones showed no protection against staurosporine-induced release of cytochrome c as determined by the percentage of cells possessing a reduced mitochondrial cytochrome c content. Statistical analysis using the

5.1 Role of heat shock protein 70 in apopotosis

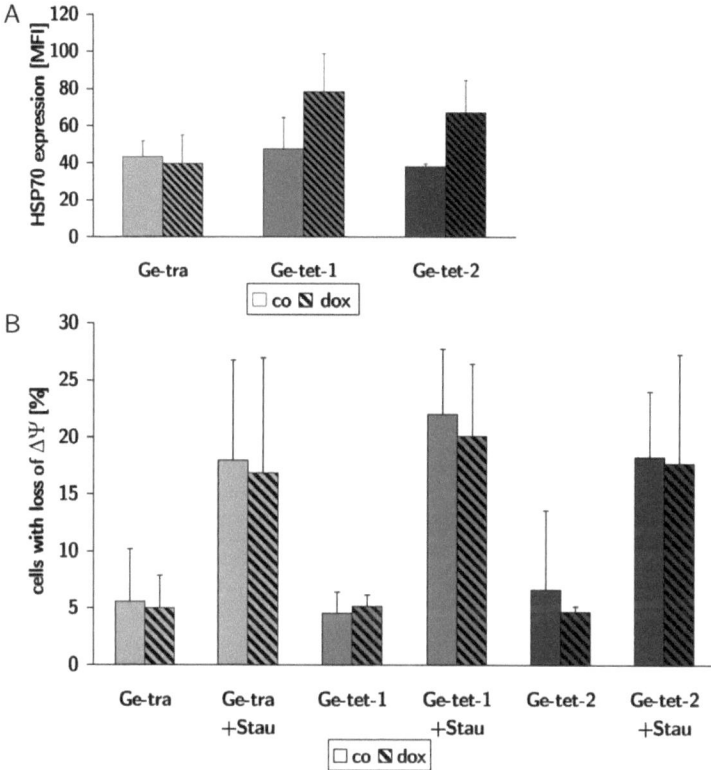

Figure 5.18: Acute HSP70 overexpression did not change the mitochondrial membrane potential $\Delta\Psi$ after staurosporine-induced apoptosis Acute induction of HSP70 in Ge-tet clones and staurosporine-induced change of $\Delta\Psi$. **A** Confirmation of acute HSP70 overexpression in the Ge-tet clones after addition of doxycycline (dox). Control (co) cells were left untreated **B** 1 μM staurosporine (+Stau) was added for 20 hrs for induction of apoptosis. As a control cells were left untreated. Cells were harvested, stained with JC-1 and fixed for flow cytometric measurement. Shown is the percentage of cells with reduced $\Delta\Psi$. Depicted are the mean values of 4 independent experiments with error bars indicating SD.

Wilcoxon-test did not show a significant decrease for the Ge-tet-1 clone (p = 0.60). The control clone Ge-tra (p = 1.00) and the other inducible clone Ge-tet-2 (p = 0.60) showed no difference in the percentage of cells with reduced cytochrome c content upon the addition of doxycycline. Thus, the acute overexpression of HSP70 in the Ge-tet-1 clone showed no significant protection of cells against staurosporine-induced mitochondrial cytochrome c release but a tendency to increase GrB-induced mitochondrial cytochrome c release.

5 Results

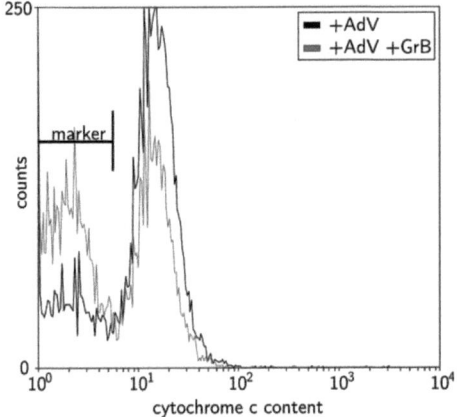

Figure 5.19: Flow cytometric analysis of the release of cytochrome c from mitochondria Apoptosis was induced by the addition of AdV and 1 ng/μl human GrB (+AdV +GrB), or as control, just the AdV (+AdV) was added for 24 hrs. Cells were fixed and stained with a cytochrome c-specific antibody and subsequently analysed by flow cytometry. For the evaluation of the data, the percentage of cells with a lower content of mitochondrial cytochrome c (under the marker) were summarised.

5.1.4.3 Effect of acute HSP70 overexpression on activation of initiator caspase-8

One of the key steps in the extrinsic apoptotic pathway is the activation of the initiator caspase-8. It can either directly activate the effector caspases-3, -6, or -7, or it can cleave BID, which then translocates to the mitochondria and with this induces the intrinsic mitochondrial pathway.

To analyse whether the acute overexpression of HSP70 had an influence on the expression levels of caspase-8 as it was indicated initially by RNA expression profiling, the following experiment was performed. Ge-tra and Ge-tet cells were treated with doxycycline for 24 hrs and subsequently apoptosis was induced using staurosporine for 20 hrs. Cellular lysates were made and probed in an immunoblot with an antibody specific for caspase-8. An anti-HSC70 antibody was used as a loading control and HeLa cells as positive control for caspase-8 detection (figure 5.22 on page 93).

Neither the acute overexpression of HSP70, nor the addition of staurosporine markedly changed the expression of caspase-8 in Ge-tet and Ge-tra cells. The upper caspase-8 band at 57 kDa represents the full-length inactive pro-caspase-8, however it is not clear, why a double-band was present in Ge and Hela cells. Low expression of the intermediate inactive cleavage form of caspase-8 with a size of 43 kDa also seems to be similar independent of HSP70 overexpression or staurosporine addition. Thus, no activation of caspase-8 on the protein level upon addition of staurosporine was observed as expected (López and Ferrer 2000). Therefore, no experiments with GrB were performed.

5.1 Role of heat shock protein 70 in apoptosis

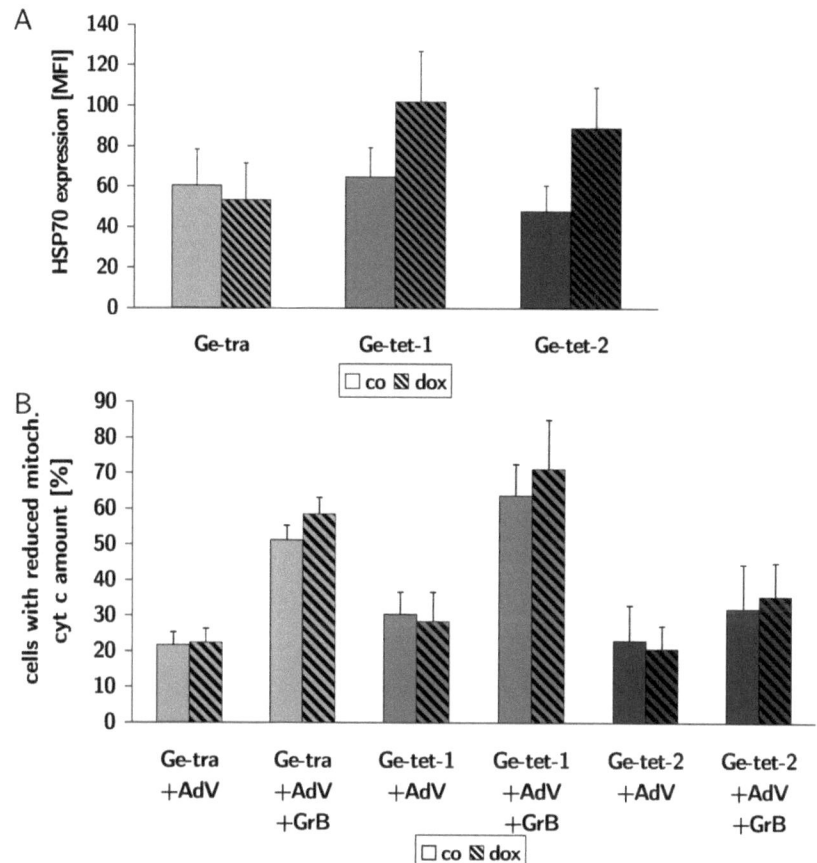

Figure 5.20: **Acute HSP70 overexpression does not significantly increase the release of cytochrome c from mitochondria after GrB-induced apoptosis** Acute induction of HSP70 in Ge-tet clones and GrB-induced apoptosis in these cells with flow cytometric measurement of mitochondrial cytochrome c release afterwards. **A** Confirmation of acute HSP70 overexpression in the Ge-tet clones. Cells were either treated with doxycycline (dox) for 24 hrs or left untreated (co). **B** AdV-β-gal and 1 ng/μl human GrB were added for induction of apoptosis. Cells were harvested, fixed, and stained for intracellular flow cytometric analysis with an anti-cytochrome c antibody and a FITC-conjugated secondary one. Depicted is the amount of cells, which shows a lower mitochondrial (mitoch.) cytochrome c (cyt c) content upon induction of apoptosis by GrB and AdV in comparison to living cells. Shown are the mean values of 6 independent experiments of Ge-tra and Ge-tet-2 and 7 independent experiments of Ge-tet-1 with error bars indicating SD.

5.1.4.4 Effect of acute HSP70 overexpression on activation of effector caspase-3

The initiator caspase-8 can directly activate the effector caspase-3 by proteolytic cleavage. Caspase-3 can also be activated by factors of the intrinsic apoptotic pathway. Due to their

5 Results

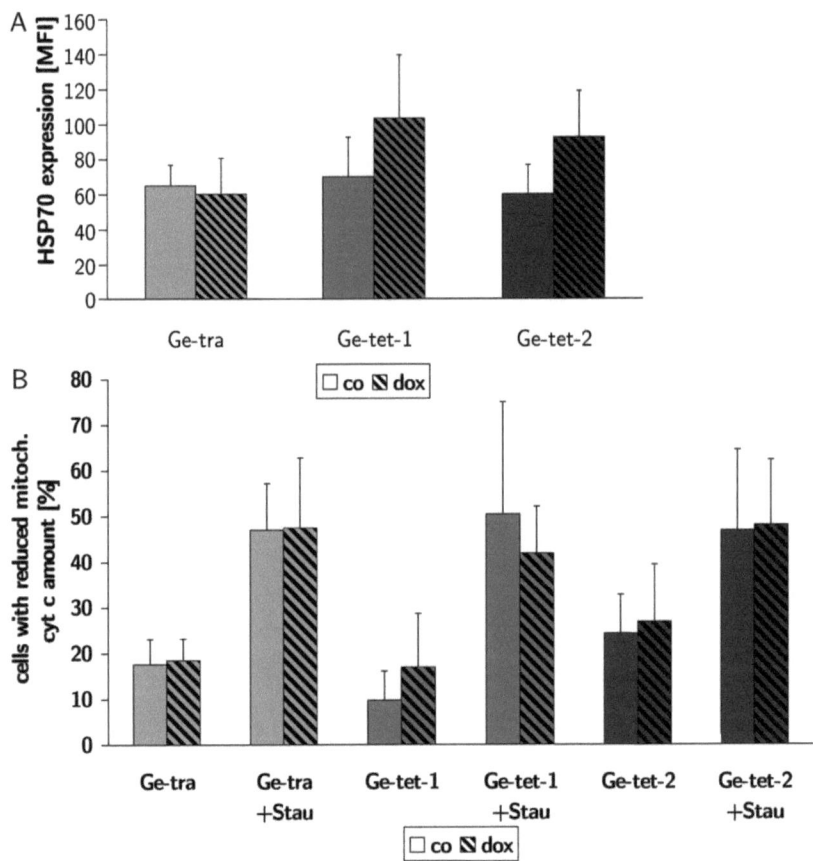

Figure 5.21: Acute HSP70 overexpression does not protect Ge-tet-1 cells from the release of cytochrome c from mitochondria after staurosporine-induced apoptosis Acute induction of HSP70 in Ge-tet clones and staurosporine-induced apoptosis in these cells with flow cytometric measurement of cytochrome c release afterwards. **A** Confirmation of acute HSP70 overexpression in the Ge-tet clones. Cells were either treated with doxycycline (dox) for 24 hrs or left untreated (co). **B** 1 µM staurosporine (+Stau) was added for 20 hrs for induction of apoptosis. Cells were harvested, fixed, and stained for intracellular flow cytometric analysis with an anti-cytochrome c antibody and a FITC-conjugated secondary one. Depicted is the amount of cells, which shows a lower mitochondrial cytochrome c (cyt c) content upon induction of apoptosis by staurosporine in comparison to living cells. Shown are the mean values of 7 independent experiments of Ge-tra and 6 independent experiments of Ge-tet-1 and 2 with error bars indicating SD.

nuclear localisation sequence ICAD and CAD translocate into the nucleus (Neimanis et al. 2007) where CAD is released from its inhibitor ICAD by caspase-3 and CAD then induces DNA fragmentation. The flow cytometric analysis of caspase-3 is based on an increase in the

5.1 Role of heat shock protein 70 in apopotosis

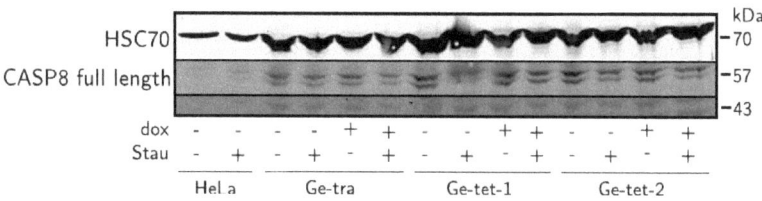

Figure 5.22: Staurosporine-induced apoptosis did not change caspase-8 levels HSP70 was acutely overexpressed in Ge-tet-1 and 2 cells by the addition of doxycycline (dox) for 24 hrs. Apoptosis was induced using 1 μM staurosporine (Stau) for 20 hrs. Cell lysates were analysed on an immunoblot using a caspase-8-specific antibody and for the loading control an HSC70-specific antibody. A plus (+) indicates the addition and a minus (-) the absence of doxycycline or staurosporine.

percentage of cells positive for active caspase-3. To determine, whether the acute overexpression of HSP70 has an effect on GrB-induced caspase-3 activation, cells were treated with doxycycline for 24 hrs and apoptosis was induced by the addition of GrB. The activation of the effector caspase-3 was measured by intracellular flow cytometry with an antibody just detecting the active cleaved form of caspase-3 (figure 5.23 on the next page). The experiments for the activation of caspase-3 were performed with the same cells as the experiments for the release of cytochrome c and therefore the induction of HSP70 was identical and is visualised in part A of figure 5.20 on page 91.

No difference in the activation of caspase-3 after the acute overexpression of HSP70 was found in GrB-induced apoptosis. It must be noted, that the overall activation of caspase-3 was low in the Ge-tet-2 clone, in contrast to the Ge-tra and the Ge-tet-1 clone and that the SD were quiet high in the Ge-tra and the Ge-tet-1 clone.

The activation of caspase-3 was also determined after staurosporine-induced apoptosis. Ge-tra and Ge-tet cells were treated with doxycycline for 24 hrs to induce HSP70 in both Ge-tet clones (part A of figure 5.24 on page 95). Apoptosis was induced by the addition of staurosporine for 20 hrs. The percentage of cells with activated caspase-3 was determined by intracellular flow cytometry (part B of figure 5.24 on page 95).

Strikingly, the acute overexpression of HSP70 reduced the percentage of cells with activated caspase-3 in Ge-tet-1 cells significantly. Statistical analysis using the Wilcoxon-test determined a significant p-value of 0.01 for Ge-tet-1. The percentage of cells with activated caspase-3 was unchanged in Ge-tet-2 cells after staurosporine-induced apoptosis ($p = 1.00$). However, also in Ge-tra cells, the percentage of cells with active caspase-3 was significantly decreased upon addition of doxycycline ($p = 0.01$).

Caspase-3 is the key player in apoptosis, as it can be activated by the extrinsic and the intrinsic pathway and can induce DNA fragmentation through cleavage of ICAD, and therefore its activation was further analysed. GrB is just one component of the cytotoxic granules of CTLs and NK cells and we wanted to investigate whether apoptosis induced by cytotoxic cells

5 Results

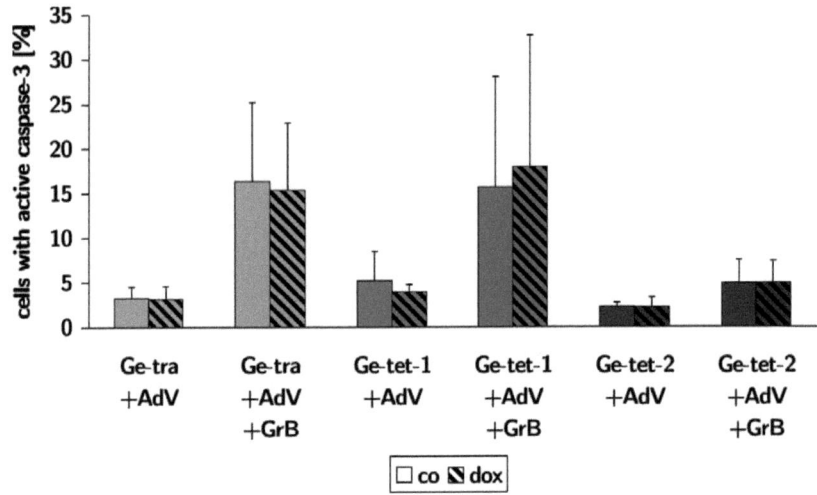

Figure 5.23: Acute HSP70 overexpression did not influence the activation of caspase-3 by GrB-induced apoptosis The acute overexpression of HSP70 was induced in Ge-tet clones upon the addition of doxycyline (dox) for 24 hrs as determined by intracellular flow cytometry. Control cells (co) were left untreated. Apoptosis was induced using 1 ng/µl human GrB and AdV-β-gal (+AdV +GrB), or as a control just the AdV-β-gal was added (+AdV). Cells were harvested, fixed, and analysed by intracellular flow cytometry using an antibody specific for caspase-3. 5 independent experiments were performed with Ge-tra and Ge-tet-1 and 4 independent experiments with Ge-tet-2 cells. Error bars indicate SD.

would affect the percentage of cells with active caspase-3.

Apoptosis was induced by co-incubation of Ge cells with NK cells. To distinguish the activation of caspase-3 in tumour cells from the one in NK cells, tumour cells were stained with DiD beforehand. By evaluating the activation of caspase-3 only in DiD-positive cells (part C of figure 5.25 on page 96), the percentage of tumour cells with active caspase-3 could be determined (part D of figure 5.25 on page 96).

It was examined, whether NK cell-mediated activation of caspase-3 was influenced by the acute overexpression of HSP70. Human NK cells were generated from whole blood and stimulated for 4 days with IL-2. Ge-tra and Ge-tet-1 cells were treated with doxycycline for 24 hrs. The acute overexpression of HSP70 in Ge-tet-1 cells was confirmed (part A of figure 5.26 on page 97). Both Ge clones were stained with DiD to be able to distinguish them from unstained NK cells in flow cytometric analysis with an antibody specific for activated caspase-3 (part B of figure 5.26 on page 97). For measuring the activation of caspase-3, killer and target cells were co-incubated at a ratio of 5:1 for 4 hrs, or target cells were incubated in the absence of NK cells (0:1).

The results of the activation of caspase-3 after NK cell-induced apoptosis showed for both clones, Ge-tra and Ge-tet-1, at both ratios, 0:1 and 5:1, that the addition of doxycycline not

5.1 Role of heat shock protein 70 in apopotosis

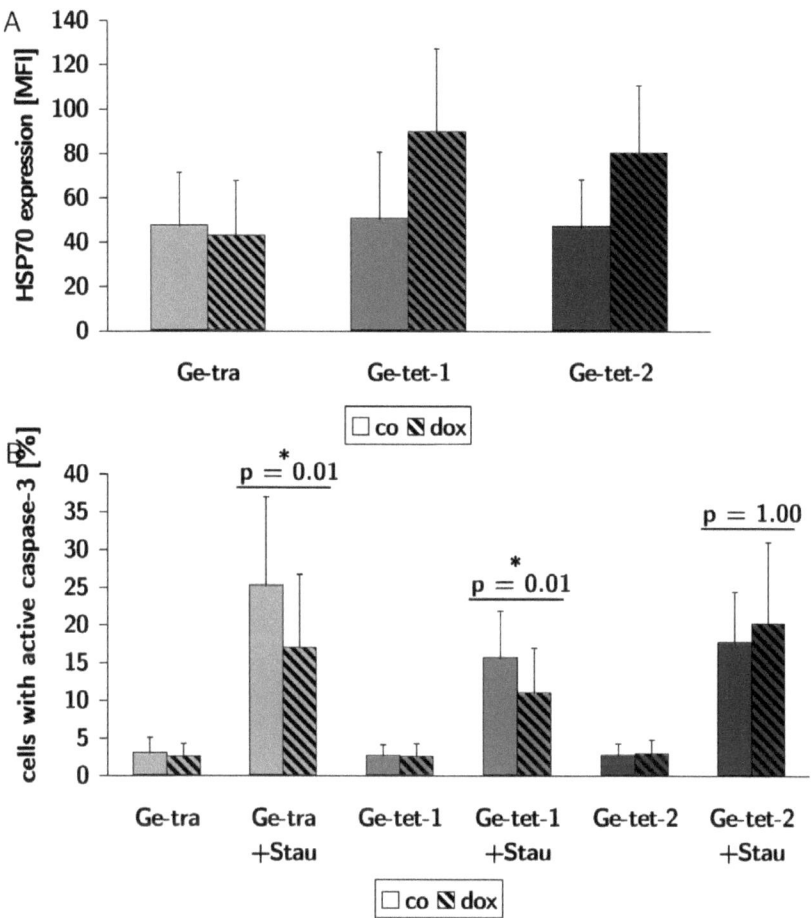

Figure 5.24: **Acute HSP70 overexpression significantly decreased the percentage of cells with active caspase-3 by staurosporine-induced apoptosis** Cells were treated with doxycycline and subsequently apoptosis was induced by the addition of staurosporine **A** The acute overexpression of HSP70 was induced in Ge-tet clones upon the addition of doxycycline (dox) for 24 hrs as determined by intracellular flow cytometry. Control cells (co) were left untreated. **B** Apoptosis was induced with 1 μM staurosporine (+Stau) for 20 hrs. The percentage of cells with activated caspase-3 was determined by intracellular flow cytometry with an antibody specific for the activated form of caspase-3. 14 independent experiments with Ge-tra, 10 independent experiments with Ge-tet-1 and 11 independent experiments with Ge-tet-2 cells contributed to the mean values shown. Error bars indicate SD. Statistical analysis using the Wilcoxon-test and an α of 0.05 was performed. An asterisk (*) indicates a significant p-value.

markedly altered the activation of caspase-3. A statistical analysis could not be done due to the small number of experiments performed.

5 Results

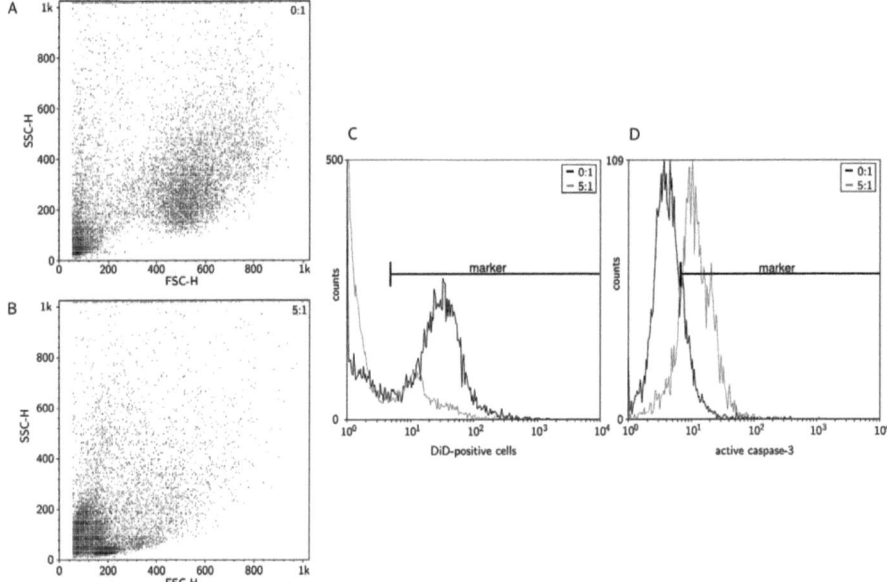

Figure 5.25: Flow cytometric analysis of the activation of caspase-3 in NK cell-mediated apoptosis Target cells were stained with DiD before apoptosis was induced by NK cells. **A** Either target cells were incubated in the absence of NK cells (0:1), or **B** NK cells were added in an effector-target ratio of 5:1 (5:1). **C** A marker was set on all DiD-positive cells and was applied as a gate to the analysis of active caspase-3 **D** The percentage of DiD-positive cells with active caspase-3 is shown.

Thus, no strong effects of the acute overexpression of HSP70 could be detected on activation of caspase-3 by different means. There rather seemed to be a tendency implying that the effect of acute HSP70 overexpression seemed to be opposite in GrB and staurosporine-induced apoptosis in the Ge-tet-1 clone. This effect did not seem to be important for NK cell-mediated activation of caspase-3.

5.1.4.5 Effect of acute HSP70 overexpression on DNA fragmentation analysed by a-poptotic ladder

DNA fragmentation is one of the main characteristics of late stage apoptosis. The DNA fragmentation was already analysed after GrB and staurosporine-induced apoptosis by sub G1-peak measurements in flow cytometry. DNA is cleaved by endonucleases like CAD, which can be directly activated by caspase-3 cleaving the inhibitor of CAD. The formation of an apoptotic DNA ladder can be visualised on an agarose gel.

It was tested, whether the acute overexpression of HSP70 had an effect on DNA laddering after GrB and staurosporine-induced apoptosis. For this purpose, Ge-tra and Ge-tet cells were

5.1 Role of heat shock protein 70 in apopotosis

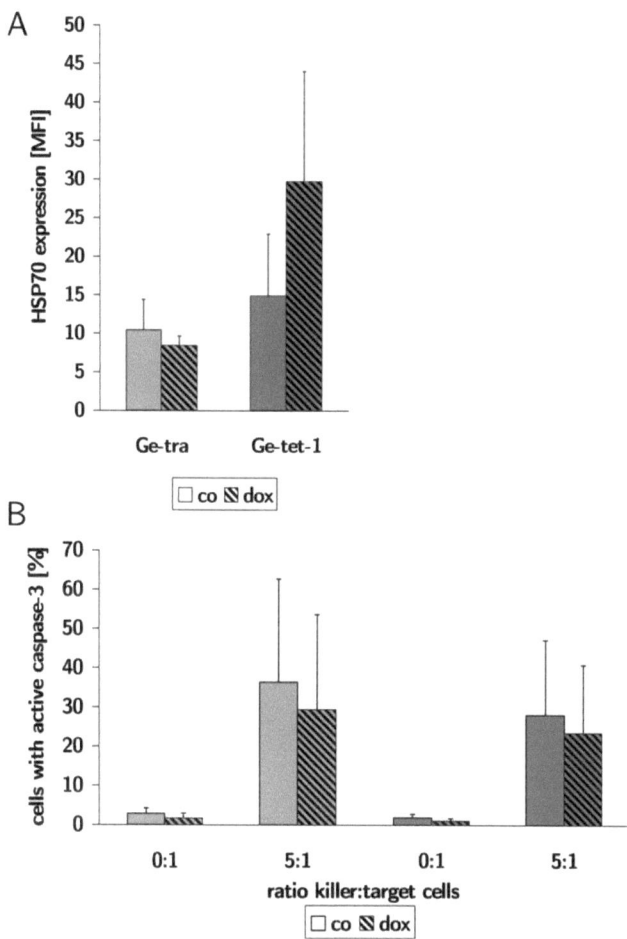

Figure 5.26: Acute HSP70 overexpression did not change the activation of caspase-3 by NK cell-induced apoptosis Acute induction of HSP70 in Ge-tet-1 cells and NK cell-induced apoptosis in these cells with flow cytometric measurement of activation of caspase-3. 4 independent experiments were performed with error bars indicating SD. **A** Confirmation of acute HSP70 overexpression in the Ge-tet-1 clone after addition of doxycycline (dox). Control (co) cells were left untreated. **B** Human NK cells and DiD-stained target cells treated with doxycycline (dox) or as control (co) left untreated were incubated at a ratio of 5:1. As a control, target cells were incubated without killer cells (ratio of 0:1). After 4 hrs of (co-)incubation the activation of caspase-3 in target cells was analysed by intracellular flow cytometry by gating on DiD-stained cells using an antibody specific for the activated form of caspase-3.

treated with doxycycline for 24 hrs and apoptosis was induced by the addition of staurosporine or GrB and AdV. As a positive control, rat Y3 cells were heat shocked for 30 min at 44 °C and

5 Results

recovered for 3 hrs at 37 °C, as they are known to show apoptotic laddering (Stark 1995). After lysing the cells and treating them with proteinase and RNase, the samples were loaded onto an agarose gel containing ethidium bromide and were visualised under UV-light (figure 5.27).

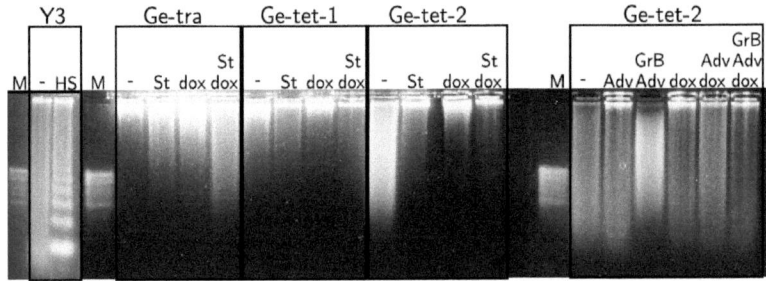

Figure 5.27: Acute HSP70 overexpression did not induce DNA fragmentation as detected by laddering of DNA Ge clones were first treated with doxycycline (dox) for 24 hrs, then apoptosis was induced either with 1 μM staurosporine (St) for 20 hrs or with 1 ng/μl GrB (GrB) and AdV (AdV) for 24 hrs. As a control AdV alone (AdV) was added. A positive control for DNA laddering were rat Y3 cells. They were heat shocked (HS) for 30 min at 44 °C and recovered for 3 hrs at 37 °C. The marker (M) is shown as well. The DNA fragmentation was visualised on an 2 % agarose gel.

In Ge cells the acute overexpression of HSP70 did not cause typical DNA laddering, neither in staurosporine nor in GrB-induced apoptosis. Only the heat-shocked positive control Y3 showed the characteristic apoptotic ladder.

In summary, there was no change after the acute overexpression of HSP70 in the loss of $\Delta\Psi$ in GrB-induced apoptosis as well as staurosporine-induced apoptosis in the intrinsic apoptotic pathway. There was a tendency of increase in cells with low mitochondrial cytochrome c content after GrB-induced apoptosis in Ge-tet-1, whereas with staurosporine-induced apoptosis, there was no effect of the acute overexpression of HSP70 in the same clone. The acute overexpression of HSP70 did not change the levels of caspase-8. The levels of active caspase-3 were also not changed by the acute overexpression of HSP70 in GrB-induced apoptosis. The addition of doxycycline to Ge-tra and Ge-tet-1 cells in combination with apoptosis induced by staurosporine decreased the percentage of cells with active caspase-3 as determined by flow cytometry.

5.2 Role of sulphatases 1 and 2 in apoptosis

How GrB is taken-up into target cells is still a matter of discussion. Several models exist, which propose different mechanisms as described in section 2.2 on page 9.

Recently, it has been shown that suppression of sulphation of HS results in a 2.6–fold reduction in the uptake of free or serglycin-complexed GrB in chinese hamster ovary (CHO)-K1 or Jurkat cells and that a reduced cell surface HS proteoglycan content leads to diminished

5.2 Role of sulphatases 1 and 2 in apoptosis

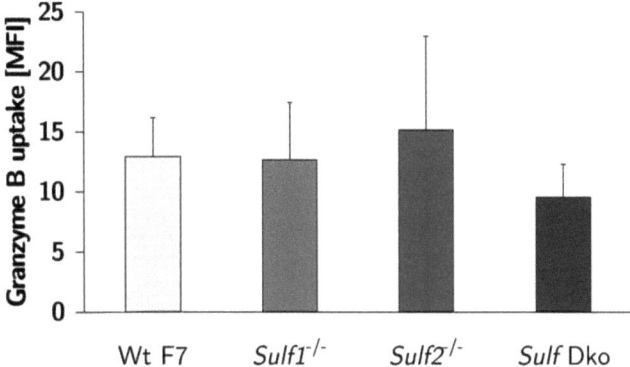

Figure 5.28: There is no difference in internalisation of GrB between *Sulf* knock-out and Wt cells Alexa 488-labelled GrB was added to the different *Sulf* clones for 1 hr before flow cytometric analysis. The uptake of GrB is shown as specific mean fluorescence intensity (MFI). The auto-fluorescence of the control was subtracted from the samples with addition of Alexa488-GrB. The results of 3 independent assays with error bars indicating the SD are depicted.

rates of GrB-induced apoptosis (Raja et al. 2005). We tested now, whether different sulphation patterns of HS affect the uptake of GrB into target cells. For this purpose MEFs deficient for HS 6-*O*-endosulphatases 1 ($Sulf1^{-/-}$) or 2 ($Sulf2^{-/-}$) or both (*Sulf* double knock-out (Dko)) were used.

5.2.1 Uptake of granzyme B into *Sulf* mouse embryonic fibroblasts

To investigate, whether the deficiency of *Sulf1* and *Sulf2* influences the uptake of GrB into MEFs Alexa-488 labelled GrB was added to the different MEFs deficient for either one or both *Sulf*s. After 1 hr of incubation the cells were harvested and washed with trypsin, to remove bound but not (yet) internalised GrB. The internalisation of GrB into viable cells was determined by flow cytometry. In figure 5.28 it is shown that the GrB uptake was approximately the same in all *Sulf* clones. These findings suggest that the knock-out of *Sulf1* and *Sulf2* does not impair the uptake of GrB into target cells.

5.2.2 Effect of the deficiency of *Sulf1* and *Sulf2* in target cells on lysis by cytotoxic T-lymphocytes

Next, we asked whether the deficiency of *Sulf1* and *Sulf2* might have a functional effect on CTL-mediated cell death. Therefore, [51]Chromium-release assays were performed and the cytotoxicity of antigen-specific CTLs on the MEFs was tested. Part A of the figures 5.29 on page 101, 5.30 on page 102, 5.31 on page 103 always shows one representative [51]Chromium-release assay with

5 Results

the positive control cell line RMA and controls for antigen-specific and calcium-dependent lysis, whereas part B shows a summary of peptide-dependent CTL-killing.

Sulf Dko cells were lysed slightly better by CTLs derived from TCR-transgenic OT-I mice than Wt cells (figure 5.29 on the facing page). RMA cells were used as positive control cell line for antigen-specific lysis by CTLs. The killing was antigen-specific as only the addition of the peptide SIINFEKL (+OVA) resulted in a lysis of the target cells, whereas its absence did prevent lysis. The addition of the calcium-inhibitor EGTA and the SIINFEKL peptide (EGTA/OVA) to the cells inhibited lysis indicating, that the CTLs used the granule-exocytosis pathway for triggering cell death. ANOVA indicated that there was a small but significant difference between the susceptibility of Wt and Sulf Dko cells towards CTLs ($p < 0.001$). To determine which gene, Sulf1 or Sulf2, was responsible for the difference, ^{51}Chromium-release assays were performed with $Sulf1^{-/-}$ and $Sulf2^{-/-}$ MEFs.

The Sulf2 gene did not seem to be responsible for this difference, as the lysis of Sulf2-deficient MEFs did not differ from that of the Wt MEFs (figure 5.30 on page 102). Statistical analysis of the data by ANOVA resulted in a p-value of 0.14, meaning that there is no significant difference.

Next it was investigated, whether instead the Sulf1 gene was responsible for the increase in lysis seen with the Dko in comparison to the Wt Sulf MEFs. $Sulf1^{-/-}$ cells showed elevated levels of lysis in comparison to Wt cells upon incubation with SIINFEKL-specific CTLs. Statistical analysis with ANOVA revealed a significant difference between the lysis behaviour of Sulf1-deficient and Wt cells with $p < 0.01$. Therefore, the small but significant difference depicted in figure 5.29 on the facing page between Sulf Dko and Wt cells is most likely caused by the knock-out of Sulf1.

5.2.3 Effect of the deficiency of *Sulf1* and *Sulf2* in target cells on apoptotic killing by cytotoxic T-lymphocytes

The increased lysis of the Sulf Dko in comparison to the Wt in the ^{51}Chromium-release assays, was found to be mainly due to the deficiency of the Sulf1 gene. This difference in cell death might be granzyme-dependent as the granule-exocytosis pathway was used. To investigate the role of granzymes we used an assay, which determines DNA fragmentation that is a sign of GrB-induced apoptosis. [^{3}H]-Thymidine is incorporated into newly synthesised DNA upon cell division. The release of [^{3}H]-Thymidine from nuclei is the result of apoptotic fragmentation of DNA into small fragments and chromatin condensation. DNA fragmentation is a hallmark of GrB-induced apoptosis (Li et al. 2001).

[^{3}H]-Thymidine-release assays were performed with Wt and Sulf double-deficient MEFs. Part A of figure 5.32 on page 105 shows a representative [^{3}H]-Thymidine-release assay with the positive control cell line RMA, and controls for peptide-specific and calcium-dependent killing,

5.2 Role of sulphatases 1 and 2 in apoptosis

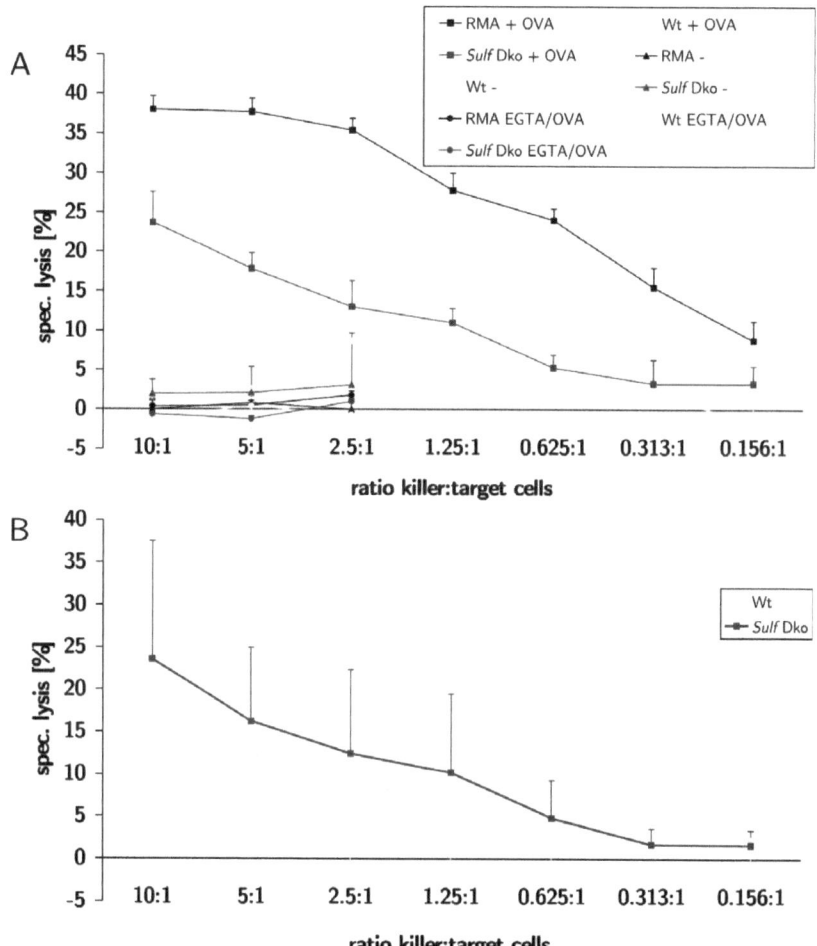

Figure 5.29: Lysis of *Sulf* Dko by OT-I-derived CTLs is slightly higher than the lysis of Wt MEFs Decreasing ratios of killer to target cells were incubated for 4 hrs. The specific lysis was calculated with formula 4.3 on page 46. **A** One representative experiment including all controls is shown. Either no peptide as negative control (-), the antigenic peptide SIINFEKL comprising aa 257 to 264 of ovalbumin (+ OVA), or the calcium-inhibitor EGTA and SIINFEKL (EGTA/OVA) were added. RMA was used as positive control cell line. Error bars indicate SD between the triplicates. **B** The mean values of 6 different assays are shown. Error bars depict SD between means of the different assays.

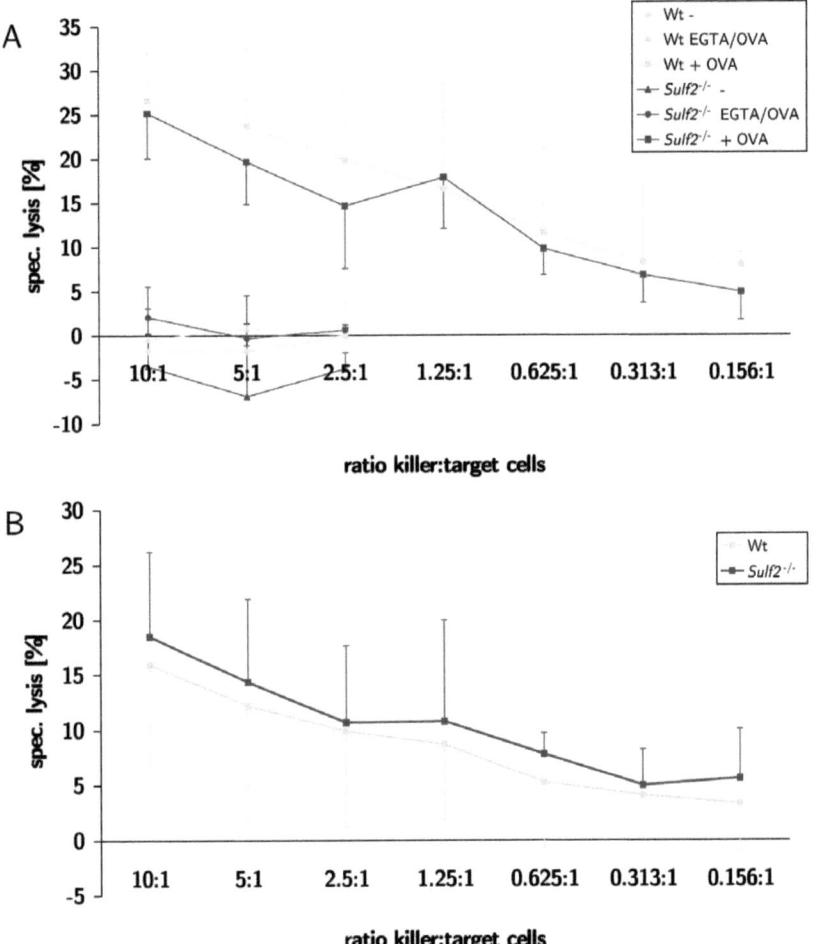

Figure 5.30: Sulf2$^{-/-}$ cells are as susceptible to antigen-specific CTLs as Wt cells. The experiments were performed as described in the legend of figure 5.29 on the preceding page. **A** One representative experiment including all controls is shown. Either no peptide as negative control (-), the antigenic peptide SIINFEKL comprising aa 257 to 264 of ovalbumin (+ OVA), or the calcium-inhibitor EGTA and SIINFEKL (EGTA/OVA) were added. RMA was used as positive control cell line. Error bars indicate SD between the triplicates. **B** The mean values of 3 different assays are shown. Error bars depict SD between the means of the different assays.

5.2 Role of sulphatases 1 and 2 in apoptosis

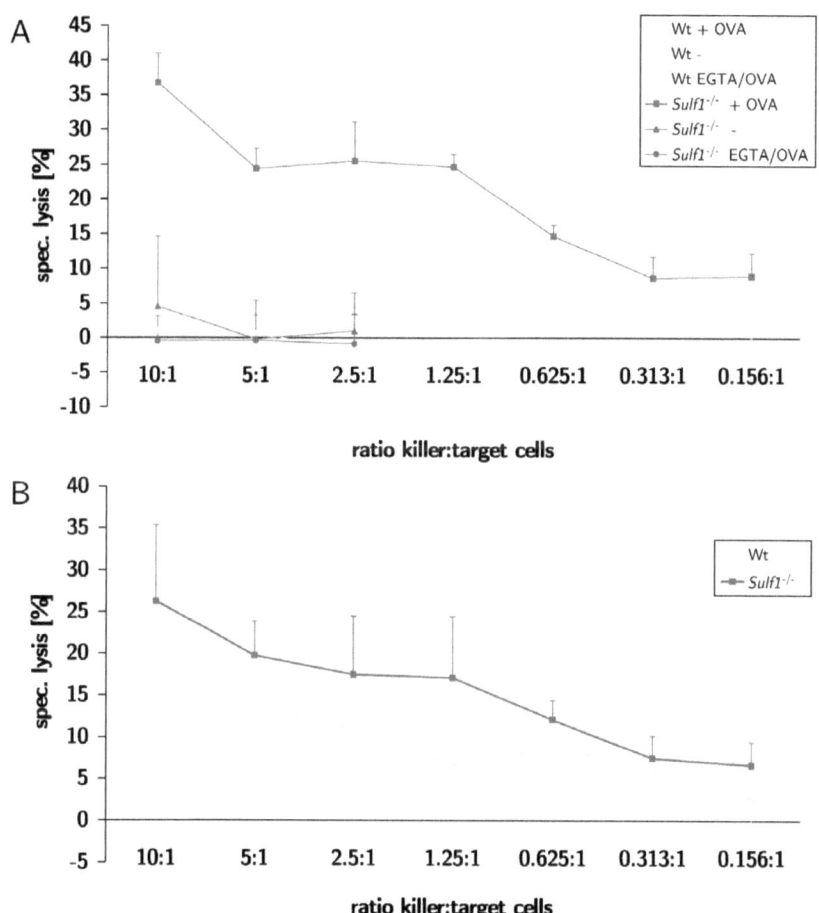

Figure 5.31: Lysis of $Sulf1^{-/-}$ cells by OT-I-derived CTLs is higher than of Wt cells Experiments were performed as described in the legend of figure 5.29 on page 101. **A** One representative experiment including all controls is shown. Either no peptide as negative control (-), the antigenic peptide SIINFEKL comprising aa 257 to 264 of ovalbumin (+ OVA), or the calcium-inhibitor EGTA and SIINFEKL (EGTA/OVA) were added. RMA was used as positive control cell line. Error bars indicate SD between the triplicates. **B** The mean values of 3 different assays are shown. Error bars depict SD between the means of the different assays.

5 Results

whereas part B shows the summary of peptide-dependent CTL-killing.

The DNA fragmentation as determined in [^{3}H]-Thymidine-release assays did not significantly differ between the double-deficient *Sulf* and Wt MEFs. The positive control cell line RMA was killed much better than the *Sulf* MEFs. The killing was antigen-specific and mediated via the granule-exocytosis pathway as determined with controls in the absence of SIINFEKL and in the presence of EGTA, respectively. The statistical analysis of the results by ANOVA also indicated no significant difference in granzyme-mediated DNA fragmentation (p = 0.17). Furthermore, the low percentage of specific [^{3}H]-Thymidine-release indicates that the DNA fragmentation of the *Sulf* MEFs was very low in general compared to the positive control cell line RMA (part A of figure 5.32 on the facing page).

The granzyme-induced DNA fragmentation of the *Sulf* single-deficient cell lines was also tested [^{3}H]-Thymidine-release assays with antigen-specific CTLs in comparison to the Wt clone.

Part A of figures 5.33 on page 106 and 5.34 on page 107 depicts a representative [^{3}H]-Thymidine-release with all controls and part B the summary of all assays.

The CTL-induced apoptosis as determined by [^{3}H]-Thymidine-release was very low and no marked difference between Wt and $Sulf2^{-/-}$ cells was detectable. Statistical analysis with ANOVA also showed no significance in DNA fragmentation between these two cell lines (p = 0.15).

Also, with *Sulf1*-deficient cells in comparison to the Wt clone, no significant difference in release of [^{3}H]-Thymidine was found. Statistical ananlysis with ANOVA determined a p-value of 0.10. The results of the experiments with both single *Sulf*-deficient cell lines indicated antigen-specificity and calcium-dependency of the killing mechanism of the CTLs.

In all experiments illustrated in figures 5.32 on the facing page, 5.33 on page 106, and 5.34 on page 107 there was no significant difference in the apoptosis rate of Wt compared to any *Sulf*-deficient cell line. However, the apoptosis rates of all MEFs were very low and therefore the chances to detect differences were rather limited. To improve the low apoptosis rates *Sulf* MEFs were treated with interferon (IFN)-γ to increase the MHC class I expression on the cell surface before conducting the [^{3}H]-Thymidine-release assays. However, this treatment did not elevate the levels of apoptosis (data not shown).

5.2.4 H2K^b expression levels on *Sulf* mouse embryonic fibroblasts

To ensure that the differences in specific lysis seen between Wt and *Sulf* Dko cells in the ^{51}Chromium-release assays were not caused by differences in the MHC class I expression on the MEFs, we determined the MHC class I expression before each experiment by flow cytometry. Higher levels of H2K^b, a mouse MHC class I molecule, that is the restriction element of SIINFEKL would allow a better presentation of the peptide to the antigen-specific CTLs and

5.2 Role of sulphatases 1 and 2 in apoptosis

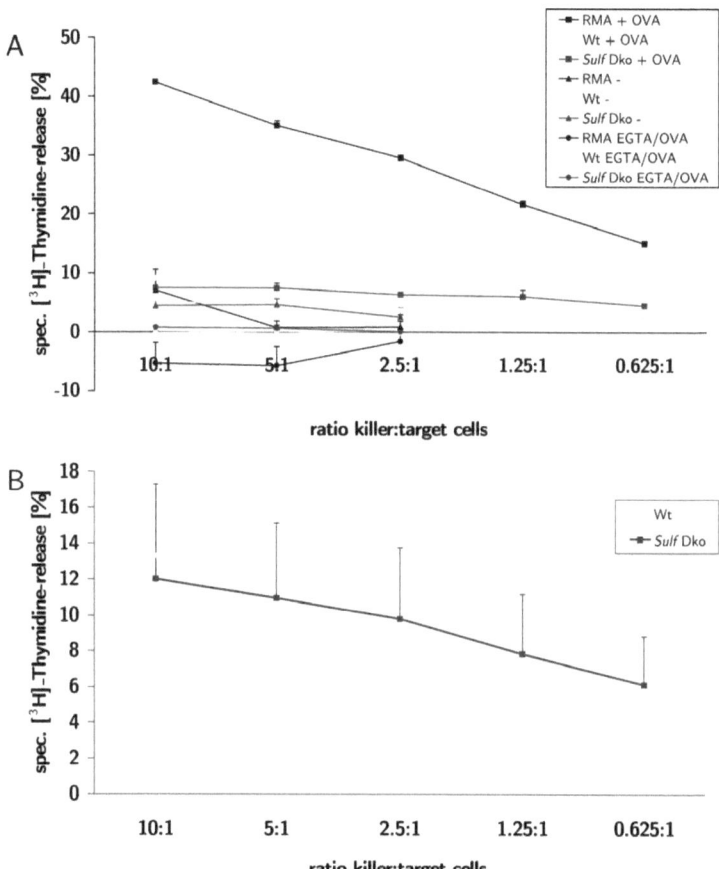

Figure 5.32: CTL-induced apoptosis does not seem to be changed in *Sulf* Dko cells compared to Wt cells
After labelling the target cells for about 20 hrs with [^{3}H]-Thymidine decreasing ratios of killer and target cells were incubated for 4 hrs. The specific [^{3}H]-Thymidine-release was calculated as described in the methods section. **A** One representative experiment including all controls is shown. Either no peptide as negative control (-), the antigenic peptide SIINFEKL comprising aa 257 to 264 of ovalbumin (+ OVA), or the calcium-inhibitor EGTA and SIINFEKL (EGTA/OVA) were added. RMA was used as positive control cell line. Error bars indicate SD between the triplicates. **B** The mean values of 5 independent assays are shown. Error bars depict SD between the means of the different assays.

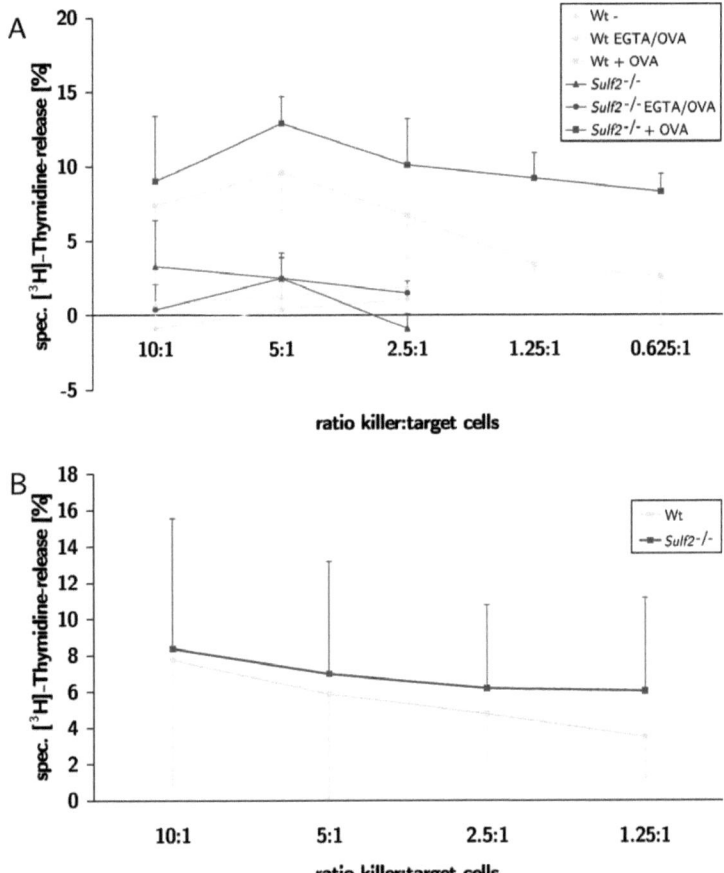

Figure 5.33: CTL-induced apoptosis is not different in $Sulf2^{-/-}$ and Wt cells The experiments were performed as described in legend of figure 5.32 on the previous page. **A** One representative experiment including all controls is shown. Either no peptide as negative control (-), the antigenic peptide SIINFEKL comprising aa 257 to 264 of ovalbumin (+ OVA), or the calcium-inhibitor EGTA and SIINFEKL (EGTA/OVA) were added. RMA was used as positive control cell line. Error bars indicate SD between the triplicates. **B** The mean values of 4 independent assays are shown. Error bars depict SD between the means of the different assays.

5.2 Role of sulphatases 1 and 2 in apoptosis

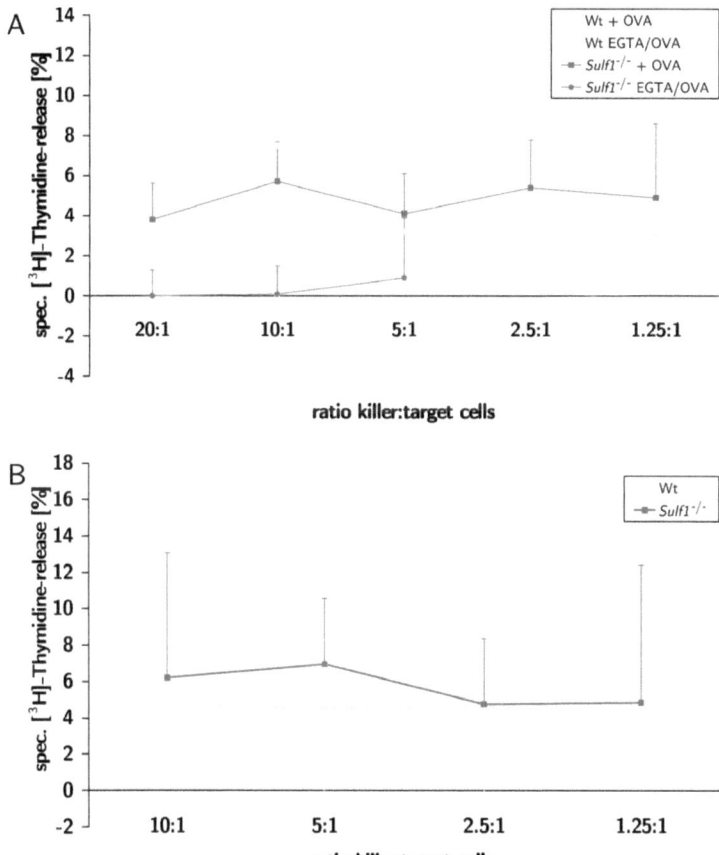

Figure 5.34: Apoptosis induced by CTLs is not different in $Sulf1^{-/-}$ cells compared to Wt MEFs A One representative experiment including all controls is shown. Either no peptide as negative control (-), the antigenic peptide SIINFEKL comprising aa 257 to 264 of ovalbumin (+ OVA), or the calcium-inhibitor EGTA and SIINFEKL (EGTA/OVA) were added. RMA was used as positive control cell line. Error bars indicate SD between the triplicates. **B** The mean values of 3 independent assays are shown. Error bars depict SD between the means of the different assays.

5 Results

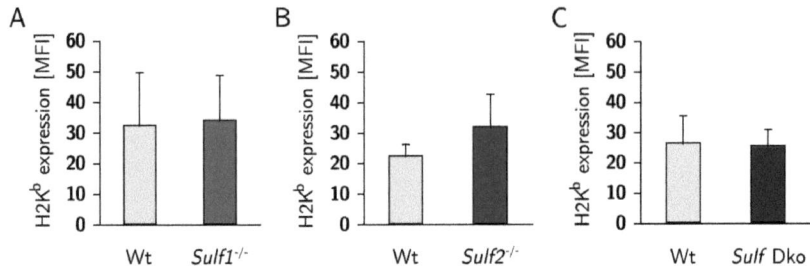

Figure 5.35: Determining H2K^b expression on Wt and *Sulf*-deficient cells H2K^b expression levels on target cells were determined by flow cytometry. Bars indicate the MFI of the H2K^b expression from which the MFI of the isotype control was subtracted. Error bars indicate SD. **A** The H2K^b expression levels of Wt and *Sulf1*$^{-/-}$ cells used in 3 ^{51}Chromium-release assays are shown. The mean values of 2 Wt and 1 Wt E9 clone contributed to the data. **B** The H2K^b expression levels of Wt and *Sulf2*$^{-/-}$ cells of 3 ^{51}Chromium-release assays are shown. **C** The H2K^b expression levels of Wt and Dko cells of 4 out of 6 ^{51}Chromium-release assays are shown. For the first two experiments no H2K^b expression levels were determined.

could therefore result in an increased specific lysis. As the H2K^b expression of *Sulf1*$^{-/-}$ cells varied in repeated experiments, two subclones of the Wt clone were generated by limiting dilution (Wt E9 and F7) with increased H2K^b expression compared to the parental clone. Depending on the H2K^b expression of the *Sulf* single or double-deficient cells in individual experiments, the Wt clone with the best matching H2K^b expression was chosen. The H2K^b expression of the *Sulf*-deficient MEFs in comparison to the different Wt clones used for all ^{51}Chromium-release assays are shown in figure 5.35. The expression of H2K^b was nearly the same after chosing the respective Wt clone so that differences determined in ^{51}Chromium-release assays were not due to a different amount of antigen-presenting molecules on the cell surface of the target cells.

For the [^{3}H]-Thymidine-release assays the Wt clone with the best matching H2K^b expression level in comparison to the *Sulf*-deficient clones was chosen as well. In contrast to the *Sulf* single-deficient cell lines, the *Sulf* double-deficient cell line showed the same H2K^b expression as the parental Wt clone. The result of the match in H2K^b expression for the different Wt clones and both *Sulf* single-deficient cell lines of all [^{3}H]-Thymidine-release assays is shown in figure 5.36 on the facing page.

5.2.5 Effect of the deficiency of *Sulf1* and *Sulf2* on apoptosis induced by granzyme B

We have demonstrated that the *Sulf1* gene has a moderate influence on lysis by CTLs. Next, we wanted to analyse, whether the deficiency of the *Sulf1* and *Sulf2* genes would influence apoptosis induced by GrB. GrB is part of the granule-exocytosis pathway, which was the pathway used by CTLs to kill *Sulf* MEFs in the cytotoxic experiments. Therefore, GrB was delivered into

5.2 Role of sulphatases 1 and 2 in apoptosis

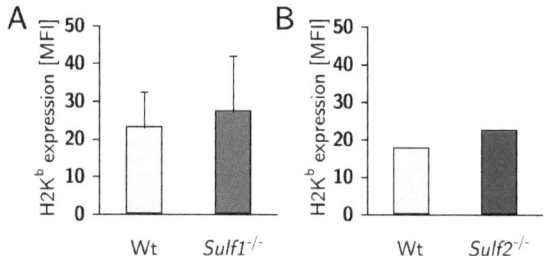

Figure 5.36: Determination of H2K^b expression on Wt and *Sulf*-deficient cells in [^{3}H]-Thymidine-release assays The H2K^b expression of the *Sulf* single-deficient cells in comparison to the Wt clone used in the [^{3}H]-Thymidine-release assays is shown. The Wt clone with the best matching H2K^b expression was always chosen. The MFI of the isotype control was subtracted from the MFI of the H2K^b antibody signal. **A** The H2K^b expression levels of Wt and *Sulf1*$^{-/-}$ cells used in 3 [^{3}H]-Thymidine-release assays are shown. The mean values of 2 Wt and 1 Wt F7 clone contributed to the data. Error bars indicate SD. **B** H2K^b expression levels of Wt and *Sulf2*$^{-/-}$ cells of 2 out of 3 [^{3}H]-Thymidine-release assays are shown. No error bars are shown as the H2K^b expression was just determined once, as the same cells were used for two [^{3}H]-Thymidine-release assays.

MEFs by replication-deficient AdV expressing a GFP protein upon transcriptional activation. The optimal MOI for *Sulf* MEFs was determined as 500 viruses per fibroblast, as tested with the Wt clone. GrB-induced apoptosis was measured by sub G1-peak measurements in flow cytometry (figure 5.37 on the next page).

The *Sulf1* and/or *Sulf2* deficiency had no clear effect on GrB-induced apoptosis. There appeared to be a tendency that the deficiency of the *Sulf2* gene and of both *Sulf* genes together decreased the GrB-induced apoptosis. However, statistical analysis using the Mann-Whitney-test indicated, that there is no significant difference in DNA fragmentation after GrB-induced apoptosis between the Wt clone and the *Sulf*-deficient clones.

5.2.6 Effect of the deficiency of *Sulf1* and *Sulf2* on transfection efficiency of adenovirus type 5

To investigate, whether differences in transfection efficiency between the single *Sulf* clones might occur, AdV-GFP was added in different MOI to all *Sulf* clones for 24 hrs (figure 5.38 on page 111). The successful production of GFP in the *Sulf* MEFs as determined by flow cytometry indicated a productive infection with the AdV.

Wt cells produced less GFP upon infection with AdV-GFP in comparison to the single *Sulf*-deficient clones. The *Sulf* double-deficient clone produced the highest amount of GFP in one set of *Sulf* MEFs. Statistical analysis with a Kruskal-Wallis-test indicated a significant overall difference between the cell lines ($p < 0.01$). Subgroup analysis using a U-test (Mann-Whithey) showed a significant difference between Wt and *Sulf1*$^{-/-}$ cells ($p = 0.02$), between Wt and

5 Results

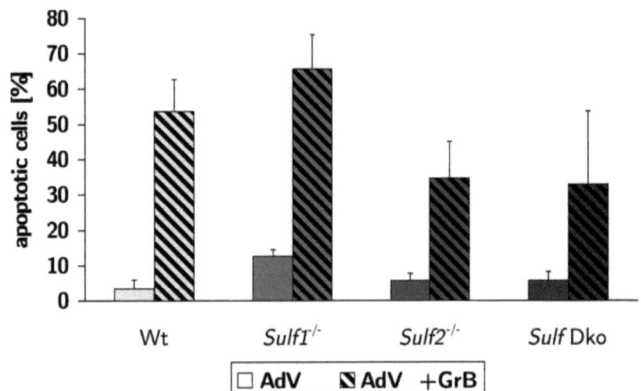

Figure 5.37: Adenoviral delivery of GrB into MEFs shows slightly reduced levels of apoptosis in the $Sulf2^{-/-}$ and the $Sulf$ double-deficient cells compared to Wt cells $Sulf$ clones were incubated with GrB and QBI-AdV-GFP (MOI 500) for 3.5–4 hrs. Cells were harvested afterwards and fixed in ice-cold ethanol for at least 18 hrs. For labelling apoptotic cells PI/RNase A/PBS was added. A sub G1-peak measurement was conducted subsequently to analyse the percentage of apoptotic cells. Cells clumping together were excluded by the doublet discrimination modus of the flow cytometer. Striped bars (AdV +GrB) represent the addition of AdV and GrB, the non-shaded bars (only AdV) represent controls, where the AdV alone was added to the cells. The results of 5 independent experiments are shown. Error bars indicate SD.

$Sulf2^{-/-}$ cells (p = 0.03), and Wt and $Sulf$ double-deficient cells (p = 0.01). Furthermore, both $Sulf1^{-/-}$ and $Sulf2^{-/-}$ cells show a significant difference in GFP expression from $Sulf$ double-deficient cells (both p-values were 0.02). To exclude clonal variability, a second set of $Sulf$ MEFs was additionally used and always indicated by an asterisk (*). Here, the $Sulf$ double-deficient cells showed the second highest amount of GFP expression. Also in the second independent set of MEFs the GFP expression was increased in all $Sulf$-deficient clones in comparison to the Wt clone. This unequal transfection efficiency stands in contrast to the results of the GrB-induced apoptosis (delivered by AdV and performed with the first set of $Sulf$ MEFs), where the $Sulf2$ and the double-deficient clones showed the lowest amount of apoptotic cells. It can be assumed that the tendency towards reduction in GrB-induced apoptosis in the $Sulf2$ and the double-deficient cell line in comparison to the Wt would have been more pronounced in these two clones, if the transfection efficiency of the AdV would be the same as in the Wt clone.

5.2.7 Expression of coxsackie and adenovirus receptor and integrin α_V on the cell surface of $Sulf$ mouse embryonic fibroblasts

After finding that the transfection efficiency of the AdV type 5 was different between the single $Sulf$ clones, the expression of the receptors for AdV uptake were investigated.

5.2 Role of sulphatases 1 and 2 in apoptosis

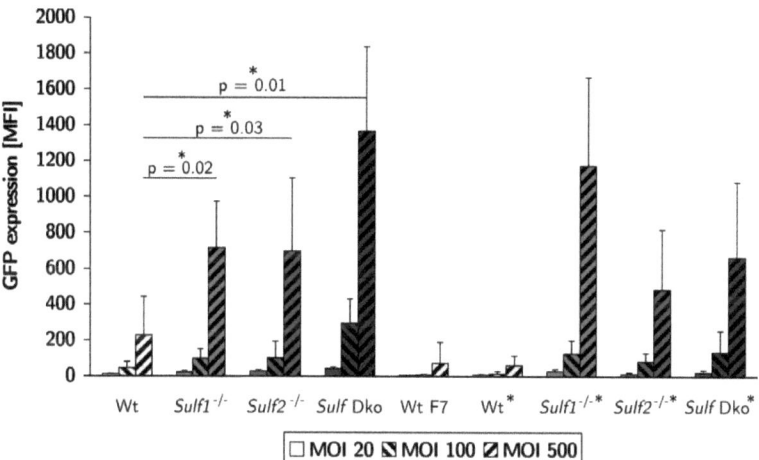

Figure 5.38: GFP expression as result of the productive infection of AdV-GFP is increased in *Sulf*-deficient clones in comparison to Wt cells AdV-GFP was added in a concentration of MOI 20, 100, and 500 to the each *Sulf* clone for 24 hrs. After harvesting, cells were fixed in 1 % PFA and viral GFP-fluorescence was measured by flow cytometry. Shown are the mean values of 7 independent assays of $Sulf1^{-/-}$ and $Sulf2^{-/-}$ cells, 6 of Wt F7 and the *Sulf* double-deficient cells and 4 of the parental Wt clone. The mean values of 3 independent experiments of all *Sulf* MEFs of the second set indicated with an asterisk (*) are shown in the right part of the figure. Statistical analysis has not been performed for the experiments with the second set of MEFs. Error bars indicate SD between the different assays. An asterisk (*) in the figure indicates a significant p-value between the MOI 500 of the different clones.

111

5 Results

As CAR is the primary receptor for binding of AdV to the cell surface (Bergelson et al. 1997; Tomko et al. 1997), we first investigated the CAR expression on all *Sulf* clones by immunoblot with a monoclonal anti-mouse CAR antibody. For loading control a monoclonal antibody against HSC70 was used, which served as standard for the subsequent densitometric analysis. The results of one immunoblot and the densitometric analysis of 3 blots are shown in figure 5.39. HeLa cells served as a positive control for CAR expression.

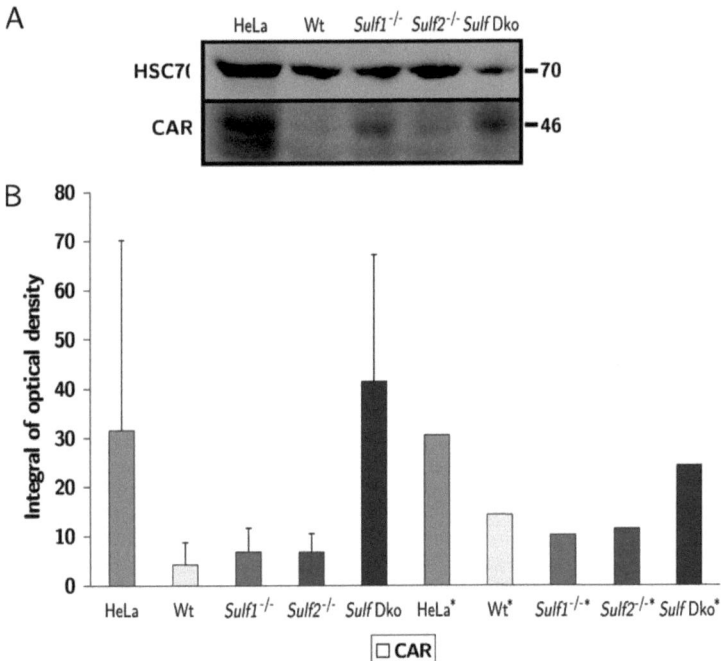

Figure 5.39: **CAR expression levels are elevated in *Sulf* Dko in comparison to Wt and *Sulf1*$^{-/-}$ and *Sulf2*$^{-/-}$ clones** CAR expression was determined by immunoblot analysis and subsequent densitometric analysis of the blots. **A** One representative immunoblot is shown. Lysates of HeLa cells and *Sulf* MEFs were made as described in section 4.3.3 on page 52. **B** The mean values of 4 densitometric analyses of 4 immunoblots are summarised in this bar chart. All clones indicated with an asterisk (*), belong to a second set of *Sulf* MEFs and HeLa as control, which were tested for CAR expression in a single experiment. The expression of HSC70 was set to 100 % (is not shown) and the relative expression of CAR was calcultated according to that. Error bars depict SD.

The expression of CAR was different in the Wt, *Sulf* single, and double-deficient clones, with the *Sulf* Dko bearing the highest CAR content as demonstrated by immunoblot and densitometric analysis. A second set of Wt, single, and double-deficient clones, indicated with an asterisk (*) showed the same pattern of CAR expression. The increased CAR expression in both *Sulf* double-deficient cell lines in comparison to the Wt and the *Sulf1*$^{-/-}$ and *Sulf2*$^{-/-}$

5.2 Role of sulphatases 1 and 2 in apoptosis

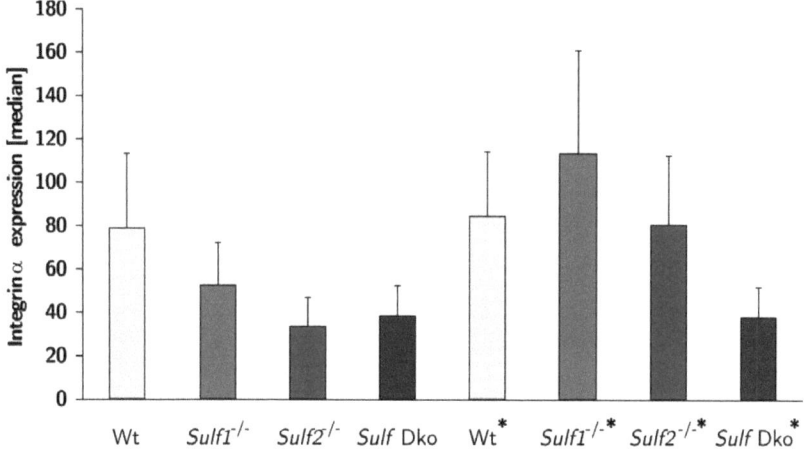

Figure 5.40: Integrin α_v content on the cell surface of the different *Sulf* clones Cells were harvested with PBS/ EDTA, washed once for 10 min at 300 x g with PBS and incubated for 45 min with anti-CD51 antibody before flow cytometric analysis. Clones labelled with an asterisk (*) belong to a second set of *Sulf* MEFs. The mean values of 4 independent experiments are summarised in this bar chart. Error bars depict SD.

cells could explain the high levels of AdV-GFP production in the Dko clones but not in the *Sulf* single-deficient ones.

Binding and uptake are two steps in AdV entry into host cells and uptake requires CAR and co-receptors, namely integrin $\alpha_v\beta_3$ and $\alpha_v\beta_5$ (Wickham et al. 1993). Thus, we investigated next, whether the integrin α_v content was different on the cell surface of the *Sulf*-deficient MEFs. For flow cytometric analysis a monoclonal antibody against the integrin α_v chain(CD51) was used (figure 5.40). The content of integrin α_v on the cell surface of *Sulf* deficient MEFs differed between the two sets of *Sulf* MEFs. In the first set of *Sulf* deficient MEFs, the integrin α_v levels were up to two times lower than in the second set (indicated with a *). Furthermore, all *Sulf*-deficient clones of the first set showed lower levels of integrin α_v compared to the Wt clone. In the second set, only the *Sulf* double-deficient clone showed lower levels of integrin α_v than the Wt. The *Sulf1*-deficient clone showed higher and the *Sulf2*-deficient clone equal levels of integrin α_v compared to the Wt clone. The constant finding between both sets was that the lowest integrin α_v level was found on the double-deficient *Sulf* clone. However, the analysis of integrin α_v levels on the cell surface also did not explain the increased GFP production of AdV-GFP in single and double *Sulf*-deficient clones.

5 Results

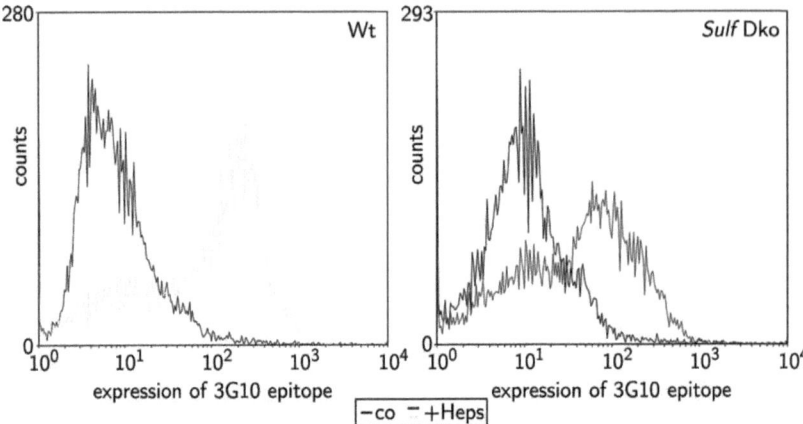

Figure 5.41: Heparinase II and III-treated *Sulf* MEFs are positive for HS stubs detected with the 3G10 antibody About 1×10^5 cells were incubated for 90 min at 37 °C with 1 mIU of enzyme per ml. Afterwards cells were washed and stained with an antibody detecting the stubs of the syndecans and glypicans and mouse anti-IgG-FITC for flow cytometric analysis.

5.2.8 Effect of heparinase II and III treatment of *Sulf* MEFs on adenoviral GFP expression

The analysis of CAR expression on all *Sulf* MEFs showed elevated levels of CAR on *Sulf* Dko cells but did not explain, the elevated GFP production of the AdV also in the single *Sulf*-deficient clones, as their CAR expression was almost the same as in the Wt clone. Next it was investigated, whether the increased GFP-production representing the increased uptake of AdV into the knock-out clones is indeed caused by the absence of the *Sulf1* and *Sulf2* genes. Due to the knock-out, the sulphation pattern of HS is changed so, that more negatively charged sulphate-groups are attached to the HS than usual (Lamanna et al. 2006). A complete removal of HS from the cell surface would diminish the uptake of AdV into the *Sulf* MEFs according to the model that HS mediate the uptake of AdV. We therefore used heparinases II and III, which were originally isolated from *Flavobacterium heparinum* (Lohse and Linhardt 1992). Heparinase II can cleave heparin and heparan sulphate-like regions of glycosaminoglycans, whereas heparinase III can just cleave heparan sulphate-like regions. The successful cleavage of HS can be tested in flow cytometry by an antibody against the stubs of the syndecans and glypicans (3G10)(Schofield et al. 1999) as it can be seen in one representative experiment in figure 5.41.

Sulf MEFs were treated with the same concentration of heparinases II and III before AdV-GFP was added for 1 hr at 4 °C to just allow for AdV binding. AdV was removed afterwards and cells were washed and incubated further 23 hrs at 37 °C. Production of viral GFP as

5.2 Role of sulphatases 1 and 2 in apoptosis

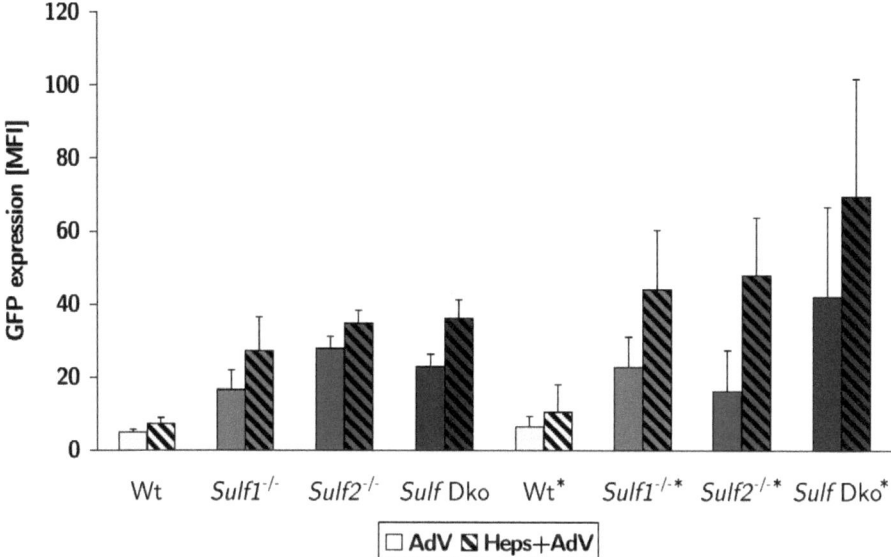

Figure 5.42: Heparinase II and III treatment of *Sulf* MEFs increases productive infection with AdV *Sulf* MEFs were treated with 1 mIU per heparinase per ml for 90 min at 37 °C, followed by 60 min incubation with AdV-GFP at 4 °C (Heps+AdV), washing and 23 hrs further incubation at 37 °C. As a control, AdV was added to the cells without heparinase treatment (AdV). AdV infection was measured in flow cytometry. The MFI of GFP after productive infection with AdV of 3 independent experiments in duplicates are shown. All clones indicated with an asterisk (*) belong to a second set of *Sulf* MEFs. Error bars depict SD.

indicator of successful infection was measured by flow cytometry as shown in figure 5.42.

Figure 5.42 demonstrates that treatment of *Sulf* MEFs with heparinases II and III even increased the uptake of AdV into these cells in both sets of *Sulf* MEFs. The increase in GFP expression was more pronounced in the *Sulf*-deficient clones in comparison to the Wt and the second set indicated with an asterisk (*) showed more pronounced differences between heparinase treated and untreated cells.

5.2.9 Heparan sulphates as receptors for type 5 adenovirus

Whereas other groups define the role of HS as co-receptors for AdV uptake (Dechecchi et al. 2001, 2000), these results of figure 5.42 challenge the co-receptor model. To investigate, whether HS really act as co-receptors for AdV uptake into MEFs, CHO cells were used. These cells naturally do not express a CAR receptor (Dechecchi et al. 2001). A745, a mutant variant of CHO cells, does not express HSPGs as confirmed by an antibody against HS entities (10E4) (figure 5.43 on the following page). With these two cell lines the uptake of AdV-GFP was

5 Results

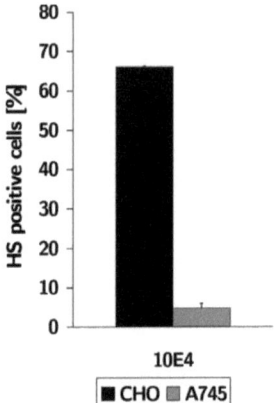

Figure 5.43: CHO mutant A745 does not express HSPGs The HS content of CHO cells and its mutant variant A745 was determined by flow cytometry with an antibody against the HS entity (10E4). The bars represent cells to which primary and secondary antibodies were added. Results of 3 independent experiments are shown. Error bars depict SD.

studied further. Figure 5.44 on the next page clarifies that the productive infection of GFP of type 5 subgroup C AdV is higher in the absence of HS, represented by the CHO mutant A745. HS therefore might not act as co-receptors for AdV uptake into MEFs but instead function as decoy receptors.

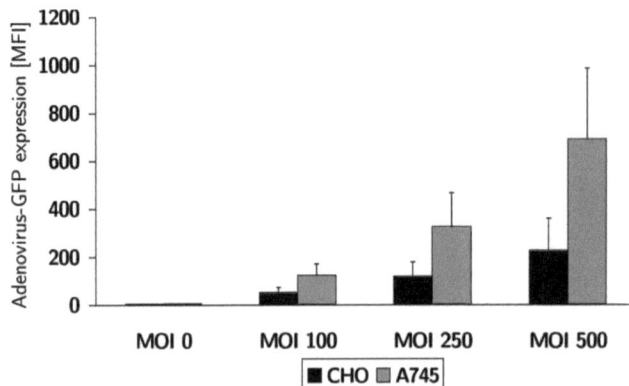

Figure 5.44: CHO mutant A745 takes up more AdV than wildtype CHO cells, which express HS AdV-GFP was added with an MOI of 100, 250, and 500 to the cells. After incubation with different amounts of virus for 24 hrs at 37 °C, cells were fixed and GFP-production was measured in flow cytometry. MOI 0 is a control with addition of no virus at all. The results of 6 independent experiments are shown. Error bars depict SD.

6 Discussion

6.1 Role of HSP70 in apoptosis

Despite the well established feature of HSP70 to have anti-apoptotic functions (Beere et al. 2000; Ravagnan et al. 2001; Saleh et al. 2000), it was demonstrated that the role of HSP70 in apoptosis seems to be far more complex as it also shows pro-apoptotic effects and is associated with a good prognosis in some tumours (Todryk et al. 1999; Trieb et al. 1998). It could earlier be demonstrated (Dressel et al. 2000, 2003, 1998) that acute but not permanent overexpression of HSP70 increases susceptibility of tumour cells towards lysis induced by CTLs as determined by ^{51}Chromium-release assays. This increase could not be attributed to an enhanced expression of MHC class I molecules on the target cell surface (Dressel et al. 2003; Dressel and Günther 1999).

6.1.1 Gene expression analysis of cells acutely overexpressing HSP70

To further elucidate whether induction of gene expression or alterations in gene transcription might be involved in the increased susceptibility of tumour cells acutely overexpressing HSP70 against CTLs, a gene expression profiling experiment was performed. RNA was isolated from the control cell line Ge-tra and from Ge-tet-1 cells acutely overexpressing HSP70 upon addition of doxycycline. The regulation of genes upon acute overexpression of HSP70 in comparison to untreated cells, and as a control, Ge-tra cells treated with doxycycline and left untreated, were compared. This experiment was not only interesting with respect to the phenotype of the cells when exposed to CTLs. In addition, it was of general interest to determine whether the overexpression of HSP70 itself can contribute to the alteration of gene expression that is observed after a stress that is associated with the induction of heat shock proteins such as HSP70.

75 genes were regulated in Ge-tet-1 cells and 43 in Ge-tra cells upon doxycycline treatment. This result is far below the usual number of genes regulated upon a certain treatment in whole human microarrays, which is around 2000 genes (Gabriela Salinas-Riester; University medicine Göttingen, Transcriptome analysis laboratory; personal communication). Except for the about 450-fold upregulation of the HSP70 gene in the Ge-tet-1 clone upon the addition of doxycycline,

6 Discussion

the other genes were maximally 23-fold upregulated (*ASNS*) in one of both clones. To obtain reliable results the microarray analysis was performed in technical duplicates and with dye swaps. The microarray procedure, beginning with the isolation of RNA and ending with the scanning of the arrays, was completely monitored and high quality was assured at all steps.

The change in expression of the genes as determined by microarray is usually validated by qRT-PCR (Provenzano and Mocellin 2007). For the qRT-PCR assays we selected *GAPDH* as housekeeping gene, as its expression was not changed upon the addition of doxycycline in both clones in the microarray. 13 candidate genes involved in apoptosis or significantly up- or down-regulated in one or both clones were chosen. For the validation of the microarray data by qRT-PCR, the same RNA as for the microarray was used. Additionally, RNA from the parental cell line Ge was used as another control and RNA from another HSP70-inducible clone, Ge-tet-2, was analysed as well. The qRT-PCR experiments revealed, that out of 13 genes, the change in expression of just one gene could be clearly confirmed qualitatively and quantitatively, which was the *Hsp70* gene, which we had induced by the addition of doxycycline in the Ge-tet-1 and the Ge-tet-2 clone. In comparison to the data of the microarray analysis the tendency of expression change was the same in 5 out of 13 genes for the Ge-tet-1 clone and in 6 out of 13 genes for the Ge-tra control clone. In the qRT-PCR analysis, out of 13 genes the tendency of expression change differed in two cases between both Ge-tet clones and 4 times between the Ge and the Ge-tra control clones, indicating that the principal tendencies of expression changes are reproducible in most but not all cases. The discrepant results between microarray and qRT-PCR could in part be caused by the fact that the RNA in the microarray analysis was transcribed using oligo dT-primer, whereas for the transcription of RNA into cDNA for the qRT-PCR random primers were used. The primers for the qRT-PCR were especially designed in such a way that the 60-mer oligos of the transcripts of the genes in the microarray were within the amplicon. Thus, it is relatively unlikely that the choice of the primers caused discrepant results. The technical execution of the qRT-PCR is also unlikely to be a reason for the discrepant expression data, as the used RNA was identical and the dissociation curves of each cDNA amplified with one primer were overlapping, indicating the same amplicon size and excluding primer mismatch. Furthermore, *Hsp70* was also found to be upregulated by doxycycline in the Ge-tet-1 clone in the qRT-PCR experiment, although not 450-fold as in the microarray but about 100-fold.

In the microarray the normalisation of the data was carried out using the fluorescence data of 44000 transcripts making it easy to set the threshold in an appropriate way. Normalisation of the qRT-PCR data was carried out by manually setting the threshold for all genes into the exponential amplification phase across all of the amplification plots and by subtracting the ct-values of *GAPDH* from the ct-values of the gene of interest. In view of the fact that the fold change in expression of *Hsp70* was about 4.5-fold higher in the microarray in comparison to the

6.1 Role of HSP70 in apoptosis

qRT-PCR it might be, that the threshold in the qRT-PCR was set too high. Thus, genes, which were significantly regulated in the microarray might not have shown up in the qRT-PCR analysis when similar criteria were applied. It could also be, as the overall number of genes regulated by the addition of doxycycline was so low in the microarray, that the percentage of false positive genes rose. Only 2 genes (*ASNS* and *CASP8*) were significantly regulated in both clones, Ge-tra and Ge-tet-1, by doxycycline addition. This suggests that most genes that appeared to be regulated by doxycycline in Ge-tra cells are false positive. Thus, also among the 75 genes that were suggested to be significantly regulated by *Hsp70* induction many false positive results might be included. In general, the results of the expression profiling experiments suggest that *Hsp70* itself does not have a major impact on gene transcription upon overexpression. This is an interesting result since it would have been possible that HSP70 as molecular chaperone could have influenced transcription factors and modified their activity as it is known for the heat shock factor (HSF), which regulates the expression of heat shock genes (Abravaya et al. 1992).

However, changes on the transcriptome level do not necessarily indicate changes on the protein level and vice versa. Thus, the effect that the acute overexpression of HSP70 increased the susceptibility of tumour cells towards CTLs is likely regulated on the protein level.

6.1.2 Effect of acute and permanent overexpression of HSP70 on early and late stages of apoptosis

To reduce the molecular complexity of CTL-mediated killing we investigated whether GrB, a protease of the cytotoxic granules of CTLs and NK cells, could be responsible for the increase in susceptibility of cells acutely overexpressing HSP70 by CTLs. For this purpose early and late stages in GrB-induced apoptosis were analysed after acute overexpression of HSP70. After acute HSP70 overexpression in the Ge-tet clones, binding of annexin V to externalised PSs was analysed in flow cytometry and DNA fragmentation caused by GrB-mediated apoptosis was analysed by sub G1-peak measurements in flow cytometry. The acute overexpression of HSP70 did not influence early stages in GrB-induced apoptosis represented by the binding of annexin V to externalised PSs. However, acute overexpression of HSP70 significantly increased the percentage of late stage apoptotic cells with fragmented DNA in the Ge-tet-2 clone ($\alpha = 0.05$; p $= 0.01$) and a similar tendency was also visible in the Ge-tet-1 clone (p $= 0.22$). Thus, acute overexpression of HSP70 does not have an anti-apoptotic effect in GrB-induced apoptosis but in contrast can have a slight pro-apoptotic effect on late stage apoptosis when apoptosis is induced by GrB. HSP70 might be able to accelerate the execution of GrB-induced apoptosis.

The sub G1-peak analysis was also performed with cells permanently overexpressing HSP70. Here, no effect of the permanent overexpression of HSP70 on GrB-induced apoptosis was de-

6 Discussion

tectable. These results are in accordance with our previous results that only the acute overexpression of HSP70 increased the susceptibility of tumourigenic cells towards CTLs (Dressel et al. 2000, 2003, 1998). This difference between acute and permanent overexpression of HSP70 was likely due to alterations in the chaperone network by long-term HSP70 overexpression, that resulted in down-regulation of HSC70 (Dressel et al. 2003).

It could be excluded that the increased percentage of apoptotic cells with fragmented DNA seen after the acute overexpression of HSP70 is caused by increased uptake of GrB into these cells. Data exist that cell surface HSP70 is able to bind and take-up GrB into tumour cells and increases NK cell-induced apoptosis (Gross et al. 2003b). However, Ge cells do not express HSP70 on the cell surface (Elsner et al. 2007, 2009).

We next analysed apoptosis in cells acutely overexpressing HSP70 that was not induced by a component of cytotoxic granules. Staurosporine, a broad spectrum protein kinase inhibitor, was chosen to induce apoptosis. Staurosporine was first isolated from the bacterium *Streptomyces staurosporeus* (Omura et al. 1977). The protein-kinase inhibitor function is achieved through its unspecific high affinity for ATP-binding sites on protein kinases with which it inhibits binding of ATP (Rüegg and Burgess 1989). In staurosporine-induced apoptosis the acute overexpression of HSP70 did not increase early stage apoptosis as determined by binding of annexin V to externalised PSs in flow cytometry. Instead, the acute overexpression of HSP70 could protect the cells, which was significant in the Ge-tet-1 clone ($p = 0.04$). Late stage apoptosis represented by DNA fragmentation and condensation as determined by sub G1-peak analysis in flow cytometry was not affected by the acute overexpression of HSP70.

In a proteomic approach, using 2-DE gel electrophoresis and mass spectrometry, it was discovered that upon staurosporine treatment two HSP70 isoforms, HSPA1B and HSPA4, HSC70 and other chaperones were upregulated in a human neuroblastoma cell line (Short et al. 2007). A significant upregulation of HSP70 in the human melanoma cell line Ge after staurosporine-induced apoptosis could not be observed (data not shown). In the literature many data exist describing a protective effect of HSP70 in staurosporine-induced apoptosis. Zhang et al. (2004) described that in human melanoma cells apoptosis induced by staurosporine involves caspase-dependent and independent pathways, whereby caspases play a major role in early stages of apoptosis and AIF in the later stages. A study in mouse macrophages revealed that upregulation of HSP70 activates BCL-XL, which then inhibits the translocation of AIF from mitochondria into the nucleus, thus, inhibiting apoptosis (Kuo et al. 2006; Ravagnan et al. 2001). It is therefore possible, that the acute overexpression of HSP70 in Ge-tet-1 cells also results in an inhibition of staurosporine-induced apoptosis by a direct or indirect inhibition of AIF by HSP70. In myoblasts, heat-shock induced overexpression of HSP70 inhibited staurosporine-induced apoptosis (Bouchentouf et al. 2004). Interestingly, HSP70 can also exert its anti-apoptotic effect in staurosporine-induced apoptosis even downstream of caspase-3 (Jäättelä et al. 1998). A protec-

6.1 Role of HSP70 in apoptosis

tive effect against staurosporine was also shown for HSP27 (Mehlen et al. 1996). Furthermore, HSP105 belonging to the HSP110 family can also suppress staurosporine-induced apoptosis by inhibiting the translocation of BAX to mitochondria (Yamagishi et al. 2006).

Next, it was analysed whether also the permanent overexpression of HSP70 would protect tumour cells from staurosporine-induced apoptosis. For this purpose, cells permanently overexpressing HSP70 or controls were treated with staurosporine and the percentage of apoptotic cells was analysed by sub G1-peak measurements in flow cytometry. As with CTL (Dressel et al. 1999) and GrB-induced apoptosis, only the acute overexpression of HSP70 but not the permanent overexpression (Dressel et al. 2003) seems to influence the outcome of apoptosis.

In summary, the acute overexpression of HSP70 does not seem to regulate the expression of many genes as determined by microarray and qRT-PCR gene expression profiling. Thus, the increased lysis of melanoma cells acutely overexpressing HSP70 by CTLs most likely is mediated on the protein level. In a first attempt to further elucidate the mechanisms behind this phenomenon, we were able to show that the late stage apoptosis induced by the cytotoxic granule effector protein GrB can be increased by acute HSP70 overexpression as well. This effect seems to be GrB-specific as the acute overexpression of HSP70 partly protected some cells against staurosporine-induced early stage apoptosis. Notably, this effect is also just seen in cells acutely overexpressing HSP70 but not in cells permanently overexpressing it. Thus, it might be that HSP70 protects or supports molecules of the pro-apoptotic pathway induced by GrB. As mentioned before, permanent and long-term overexpression of HSP70 is compensated by a down-regulation of HSC70 indicating adaptations of the chaperone network to high HSP70 levels (Dressel et al. 2003).

To elucidate, which steps in the apoptotic pathway were affected by acutely overexpressed HSP70, further experiments were performed.

6.1.3 Analysis of key steps in apoptosis after the acute overexpression of HSP70

The acute but not the permanent overexpression of HSP70 had increased the percentage of late apoptotic Ge-tet-2 cells in GrB-induced apoptosis but on the other hand also partly protected Ge-tet-1 cells from staurosporine-induced externalisation of PS. To identify key steps in apoptosis, which might be affected by the acute overexpression of HSP70 in GrB or staurosporine-induced apoptosis the following steps of the apoptotic pathways were analysed further: The loss of the mitochondrial membrane potential $\Delta\Psi$, the release of cytochrome c from mitochondria, the activation of initiator caspase-8 and effector caspase-3, and the DNA fragmentation by apoptotic laddering.

Flow cytometric analysis of the loss of $\Delta\Psi$ after induction of apoptosis revealed the follow-

6 Discussion

ing: GrB and staurosporine where both able to induce a loss of $\Delta\Psi$. The loss of $\Delta\Psi$ was slightly more pronounced in GrB-induced apoptosis in comparison to staurosporine-induced apoptosis. Neither the GrB nor the staurosporine-induced change of $\Delta\Psi$ was significantly affected by the acute overexpression of HSP70. The GrB-induced loss of $\Delta\Psi$ is in accordance with literature (Alimonti et al. 2001; MacDonald et al. 1999; Waterhouse et al. 2006a) and also the staurosporine-induced change of the mitochondrial membrane potential $\Delta\Psi$ was described before (Scarlett et al. 2000). Charlot et al. (2004) further demonstrated that the loss of $\Delta\Psi$ is an early event in staurosporine-induced apoptosis.

The second key event of the intrinsic apoptotic pathway investigated was the release of cytochrome c from mitochondria. Flow cytometric analysis using a cytochrome c-specific antibody illustrated that both GrB and staurosporine-induced apoptosis were able to increase the proportion of cells with a reduced mitochondrial cytochrome c content. This effect was again slightly more pronounced in the GrB-induced apoptosis. The effect of acute HSP70 overexpression on the release of mitochondrial cytochrome c varied in GrB and staurosporine-induced apoptosis. The acute overexpression of HSP70 in the Ge-tet-1 clone slightly increased the percentage of cells (with a borderline significance of $p = 0.06$) with a reduced mitochondrial cytochrome c content after GrB-induced apoptosis, whereas a tendency towards protection could be detected in the same clone in staurosporine-induced apoptosis. GrB has been reported to trigger the release of cytochrome c from mitochondria induced by cleaving BID, so that its truncated version tBID can then recruit BAX to the mitochondria, which forms pores and induces the release of cytochrome c (Alimonti et al. 2001; Heibein et al. 2000). Staurosporine is as well capable of inducing the release of cytochrome c from mitochondria (King et al. 2007). Furthermore, the intrinsic mitochondrial pathway seems to be the sole or at least the main pathway of staurosporine to trigger apoptosis (King et al. 2007). Thus, HSP70 might cause a stronger or faster activation of the intrinsic pathway in these cells upon GrB addition. The differences between the Ge-tet-1 and the Ge-tet-2 clones indicate that the effect of HSP70 can vary depending on the cellular environment.

One key step in the extrinsic apoptotic pathway is the activation of the initiator caspase-8. This caspase can either directly activate the effector caspase-3 by proteolytic cleavage or proteolytically cleave BID, to initiate the intrinsic pathway. A second reason why the expression and activation of caspase-8 was investigated, was its significant 22-fold down-regulation in the Ge-tet-1 clone upon doxycycline treatment as determined by microarray analysis although this was not confirmed in qRT-PCR. The expression and activation of caspase-8 was investigated by immunoblot analysis with an antibody described to detect the 57 kDa inactive zymogen, the cleaved intermediate with a size of 43 kDa, and the active caspase-8 with a size of 18 kDa. Neither the addition of doxycyline for expression of HSP70 in the Ge-tet clones, nor the addition of staurosporine to induce apoptosis, changed the expression or activation of the 57 or the 43

6.1 Role of HSP70 in apoptosis

kDa bands. The literature for an activation of caspase-8 by staurosporine is controversial. López and Ferrer (2000) demonstrated that staurosporine does not trigger a caspase-8 dependent pathway. In contrast, Nicolier et al. (2009) suggested that staurosporine induces cell death via caspase-8 or caspase-9 signalling cascades leading to the induction of the intrinsic pathway but also state that a caspase-independent pathway can be triggered. Furthermore, they claim that the involvement of caspases in triggering staurosporine-induced apoptosis is dependent on p53, whether it is present in a Wt or mutant form. Additionally, the choice, which apoptotic pathway is induced might also depend on the concentration of staurosporine used, whereby a higher concentration might trigger the activation of caspase-8 as demonstrated by Nicolier et al. (2009). They used 300 nM staurosporine and could demonstrate caspase-8 activation, whereas López and Ferrer (2000) used 100 nM and could not show activation of caspase-8. We on the other hand used even 1 μM staurosporine and could not detect an activation of caspase-8.

The most important among the key steps in apoptosis is the activation of caspase-3 as it can be activated by factors of both the intrinsic and the extrinsic pathway and mediate DNA fragmentation by releasing CAD from its inhibitor ICAD to cleave DNA. GrB, staurosporine, and NK-cells were able to trigger the proteolytic activation of caspase-3 in Ge cells. The acute overexpression of HSP70 did not have a significant effect on the activation of caspase-3 in GrB-induced apoptosis but significantly protected Ge-tet-1 cells from activation of caspase-3 in staurosporine-induced apoptosis. Strikingly, the addition of doxycycline to the control clone Ge-tra also significantly protected these cells from staurosporine-induced apoptosis. Thus, it must be questioned, whether in this case, the protection is an effect of the acute overexpression of HSP70.

In vivo analysis in human and mouse systems using an inhibitor of caspases, namely zVAD-fmk, revealed that GrB can directly proteolytically cleave caspase-3 (Atkinson et al. 1998). GrB and GrB-activated caspase-3 then synergistically activate effector caspase-7 (Yang et al. 1998). It could be demonstrated with the same inhibitor of caspases that staurosporine can activate caspase-3 as well (Chae et al. 2000; Yue et al. 1998). In some cases the activation of caspase-3 even seems to be neccessary for activation of caspase-8 and cleavage of BID to amplify weak mitochondrial signals in staurosporine-induced apoptosis in breast cancer cells (Tang et al. 2000). In hyperosmolarity-induced apoptosis, mice deficient for *Hsp70* showed a rapid activation of caspases-3 and -9, whereas HSP70 in Wt mice protected the cells from caspase-3 activation and thus apoptosis.

NK cell-mediated activation of the key molecule caspase-3 was also investigated to see whether it would make a difference to GrB-induced apoptosis. Pre-labelling of doxycycline-treated and untreated target cells with DiD allowed to distinguish the activation of caspase-3 in target and effector cells. DiD, a lipophilic dye incorporating into plasma membranes, was first used by Honig and Hume (1986) for labelling living neurons and no toxic effect of the dye was observed.

6 Discussion

Co-incubation of effector and target cells at a ratio of 5:1 increased the percentage of cells with active caspase-3 in comparison to target cells alone. The acute overexpression of HSP70 in Ge-tet-1 cells did not affect the activation of caspase-3, as it was also seen in GrB-induced apoptosis. Thus, the HSP70 effect on caspase-3 activation after GrB-induced apoptosis does not explain the increased lysis observed in the granule-exocytosis pathway with CTLs.

DNA fragmentation was analysed by sub G1-peak analysis in flow cytometry before, but we decided to analyse it again, as it seemed to be significantly increased after acute overexpression of HSP70 in GrB-induced apoptosis. This time DNA fragmentation was analysed by visualising the DNA ladder on an agarose gel. Here, the rungs of the ladder are caused by internucleosomal cleavage of DNA by endonucleases, yielding DNA fragments with a size of 180 bp or a multiple of it. Neither GrB nor staurosporine-induced apoptosis were able to induce DNA laddering. Only the positive control, namely heat-shocked Y3 cells showed the characteristic apoptotic DNA ladder. According to the literature GrB is able to induce DNA fragmentation by direct cleavage of ICAD without a requirement for caspase-3 (Sharif-Askari et al. 2001; Thomas et al. 2000). Staurosporine-induced apoptosis on the other hand is able to activate CAD through proteolytic cleavage carried out by caspase-3 (Lechardeur et al. 2000; Tang and Kidd 1998). There might be two reasons why we do not see an apoptotic ladder after staurosporine and GrB-induced apoptosis. The concentration of staurosporine used by Tang and Kidd (1998) was identical with the concentration we used but the concentration of GrB used by Thomas et al. (2000) to show apoptotic laddering was between 2.0 and 3.5 μM, whereas we used about 2.9 pM, which makes a difference in concentration of roughly 7000 to 12000-fold. Notably, they were not able to detect apoptotic laddering with concentrations of less than 0.5 μM. Secondly, the visibility of DNA laddering might also depend on the cell type.

The binding of annexin V to exposed PSs, the loss of $\Delta\Psi$, the release of cytochrome c from mitochondria, and the activation of caspase-3 were a lot less pronounced in the Ge-tet-2 clone in comparison to both other clones, Ge-tet-1 and Ge-tra, but just upon GrB-induced apoptosis. Upon staurosporine-induced apoptosis the values were similar to the other clones. All pro and anti-apoptotic tendencies or effects of the acute overexpression of HSP70 in combination with GrB and staurosporine-induced apoptosis were more pronounced in the Ge-tet-1 clone. The only exception is that GrB-induced fragmentation of DNA was significantly upregulated in the Ge-tet-2 clone. GrB is able to directly process ICAD without the requirement for caspase-3, which just amplifies the signal and MEFs deficient for *Dffa* (the gene encoding DNA fragmentation factor, 45 kDa; also known as ICAD (DFF45)) are impaired in GrB-induced apoptosis (Thomas et al. 2000). All other activation pathways normally initiated by GrB seem to be inhibited or downregulated explaining the comparibly low activation of apoptosis in comparison to Ge-tet-1 and Ge-tra as indicated by binding of annexin V to exposed PSs, the loss of $\Delta\Psi$, the release of cytochrome c from mitochondria, and the activation of caspase-3. Additionally, early

6.1 Role of HSP70 in apoptosis

stage apoptosis in the Ge-tet-2 clone seemed to be less affected by the acute overexpression of HSP70, although the induced levels of HSP70 were comparable to the ones in the Ge-tet-1 clone. Acutely overexpressed HSP70 might chaperone pro-apoptotic factors in GrB-mediated apoptosis, and GrB might circumvent other apoptotic pathways and directly activates DNA fragmentation in Ge-tet-2 cells. Thus, Ge-tet-2 cells could be less affected in intermediate apoptotic steps in comparison to Ge-tet-1 cells. Namely, if specific apoptotic pathways in Ge-tet-2 cells are inhibited or not even triggered, HSP70 cannot chaperone pro-apoptotic factors in these pathways and therefore no effect of acute overexpression of HSP70, except for the one on DNA fragmentation analysed by sub G1-peak analysis, can be seen.

The differences between the Ge-tet-1 and the Ge-tet-2 clone might be explained by clonal variations. In the qRT-PCR analysis the clones also just showed the same tendency of regulation in 11 out of 13 genes, indicating that they might differ in some respects due to spontaneous mutations.

In summary, the acute overexpression of HSP70 was able to increase DNA fragmentation as determined by sub G1-peak analysis and a tendency was observed that it also could increase the release of cytochrome c from mitochondria in GrB-induced apoptosis. In staurosporine-induced apoptosis the acute overexpression of HSP70 partly protected cells from externalisation of PS in early apoptosis and from activation of caspase-3. Thus, the acute overexpression of HSP70 seems to play a role in the intrinsic pathway, where it can either partially protect cells from staurosporine-induced apoptosis or slightly augments GrB-induced apoptosis.

The extend of overexpression of HSP70 with the Tet-On system in Ge cells was rather moderate compared to the permanent overexpression of HSP70 in the virally transduced system. Nevertheless, we found some effects in apoptosis due to acute overexpression of HSP70. In general, the effects of HSP70 in GrB-induced apoptosis are too moderate to explain the effects on CTL-induced lysis that we had observed previously (Dressel et al. 2003, 1999). Thus, other components of the granule-exocytosis pathway might be more important in this respect.

To find functionally relevant interacting partners of HSP70 in cells acutely overexpressing HSP70 also another approach was started. An HSP70 fusion protein linked to the protein transduction domain TAT of the human immunodeficiency virus-1 (HIV-1) was generated (data not shown). TAT should allow easy transfer of HSP70 into cells without stressing them but by simple addition of the fusion protein into the medium (Fawell et al. 1994). A 6-His-tag at the C-terminus of the fusion protein could allow fishing in these cells for interacting partners and identifying them with co-immunoprecipitation and mass spectrometry. Unfortunately, we could neither convincingly confirm the uptake of this fusion protein nor a similar HSP70 fusion protein (Nagel et al. 2008) into our cells by intracellular flow cytometry or by confocal microscopy.

In SCID mice (deficient for B-cells and T-cells), we were able to demonstrate that tumours derived from the human melanoma cell line Ge-Hsp70, permanently overexpressing HSP70, grew

6 Discussion

slower and did not give rise to metastases in contrast to tumours derived from the same cell line overexpressing control proteins. The partial control of tumour growth could be attributed to an augmented cytotoxic activity of NK cells in these animals. In SCID/beige mice (additionally lacking NK cells) growth of HSP70-overexpressing tumours and the frequency of metastases were not altered. The cytotoxic activity of NK cells could be stimulated by HSP70-positive exosomes that were released from the HSP70-overexpressing tumours (Elsner et al. 2007). It was demonstrated that stimulation of NK cells either with HSP70 or the HSP70-derived peptide TKD (TKDNNLLGRFELSG) but not HSC70 increased the cytotoxic activity of NK cells and their expression of GrB (Elsner et al. 2009). In this model MHC class I chain-related molecule (MIC)A and MICB molecules on target cells functioned as recognition structures for stimulated NK cells (Elsner et al. 2007, 2009). The HSP70-mediated stimulation of NK cells might either be directly mediated through C-type lectin NK receptors (Gross et al. 2003a; Thériault et al. 2006) or via HSP70 receptors on APCs and cross-talk to NK cells (Degli-Esposti and Smyth 2005).

During apoptosis, which is a controlled cell death, apoptotic cell debris is rapidly engulfed by phagocytes (Savill et al. 1993) to avoid leakage of the cytoplasmic content of the cells, which might release alarmins or danger signals (Oppenheim and Yang 2005), which could recruit cells of the immune system (Srivastava 2002a, 2002b). Heat shock proteins were described to belong to the group of danger signals and the term "chaperokine" was proposed to describe their chaperoning and cytokine-like functions (Asea et al. 2000). Danger signals are just released from apoptotic cells, when they outnumber the ability of phagocytes to engulf the apoptotic cell debris (Scaffidi et al. 2002; Shi et al. 2003). HSP70 appears to be also released therefore from apoptotic or necrotic cells and attract cells of the immune system. Thus, it is not surprising that in many animal models HSP70 expression in tumours is associated with their regression (Clark and Menoret 2001; Melcher et al. 1998; Menoret et al. 1995; Todryk et al. 1999; Wells et al. 1998).

In conclusion, the stimulatory effects of HSP70 acting as danger signal on cells of the immune system in vivo might be even more important for cancer biology than intracellular effects during apoptosis.

6.2 Role of sulphatases 1 and 2 in apoptosis

The *Sulf* MEFs are a valuable tool to investigate the role of differentially sulphated HS in CTL and GrB-induced apoptosis. During the studies it was discovered that the uptake of AdV into these cells also seemed to be strangely influenced by sulphation of HS.

6.2 Role of sulphatases 1 and 2 in apoptosis

6.2.1 Interaction of granzyme B and mouse embryonic fibroblasts deficient for *Sulf1* and *Sulf2*

The textbook model of the uptake of granzymes into target cells during CTL-mediated cytotoxicity describes that perforin forms pores in the target cell membrane through which granzymes can enter the cell. This model was challenged by the finding that granzymes are sequestered in a complexed form bound to the 250 kDa chondroitin sulphate serglycin, which is then way too large to enter through those pores (Metkar et al. 2002; Raja et al. 2002). Furthermore, sublytic concentrations of perforin, which do not form pores, are sufficient to release granzymes from endosomes and to induce apoptosis (Froelich et al. 1996a). Several other pathways for the uptake were suggested including dynamin-dependent and independent pathways (Tran et al. 2003; Veugelers et al. 2004) and pathways for which the MPR300 is essential (Motyka et al. 2000; Veugelers et al. 2006) or not (Dressel et al. 2004a, 2004b; Kurschus et al. 2005).

Raja and colleagues (Raja et al. 2005) could demonstrate that the anionic granzymes exchange from the cationic serglycin to more cationic sites on the cell surface (sulphated PGs). By using cell lines either lacking xylosyl transferase or HS polymerase, mutants with decreased levels of PGs or HSPGs, respectively, were gained. Both of them showed a diminished uptake of GrB indicating that those cell surface structures are involved in the uptake of GrB. The findings suggest that sulphation of cell surface HS plays a role in the uptake of GrB into target cells and led us to analyse the effect of the knock-out of HS 6-*O*-endosulphatases 1 and 2 on GrB uptkake. The avian QSULF1 and QSULF2 desulphate cell surface HS (Ai et al. 2006, 2003). A knock-out of mouse *Sulf1* and *Sulf2* genes leads to higher sulphated cell surface HS on MEFs (Lamanna et al. 2006).

It was tested, whether the increased sulphation of MEFs deficient for *Sulf1* or *Sulf2* or both, would lead to elevated uptake of GrB and therefore elevated levels of apoptosis. The level of GrB-uptake did not differ between the single *Sulf* clones as determined by the addition of Alexa488-labelled GrB for 60 min. Raja and colleagues (Raja et al. 2005) and also Kurschus and colleagues (Kurschus et al. 2005) detected a correlation between the density of sulphation of HS and the uptake of GrB. After treatment with sodium chlorate, which generally inhibits the sulphation of HS-chains, both groups could show a concentration-dependent reduction in GrB binding but also a 2.6-fold decrease in uptake of monomeric or complexed GrB in Jurkat and CHO-K1 cells. According to our results an increase in sulphation of HS-chains does not augment the levels of GrB binding and uptake. This does not necessarily indicate a discrepancy between our findings and literature. Sodium chlorate is a selective inhibitor of ATP sulphurylase (ATP sulphate adenylyltransferase), the first enzyme in the sulphate activation pathway, to activate sulphation. ATP sulphurylase does not only activate sulphate synthesis but also catalyses GTP-hydrolysis, both of which reactions are coupled, implying the possibility that more than

6 Discussion

just the sulphation of HSs is abrogated (Sukal and Leyh 2001). On the other hand, avian QSULF1 and QSULF2 were not only shown to specifically desulphate cell surface HS but also cell matrix HS (Ai et al. 2006). The enzyme specificity of SULF1 and SULF2 is not restricted to di- and tri-sulphated 6S disaccharide units within the HS chain, but additionally the *Sulf1* and *Sulf2* genes also have an impact on the 6O-sulphotransferase activity (Lamanna et al. 2008). Therefore, the knock-out of *Sulf1* and *Sulf2* leads to specific patterns of sulphation and not to a general suppression of sulphation as in a sodium chlorate treatment. Furthermore, it might be that a higher level of sulphation does not increase binding and uptake of GrB into cells, because a saturation could occur maybe caused by steric hindrance.

To analyse whether CTL-mediated lysis is affected by the deficiency of the *Sulf1* and *Sulf2* genes, ^{51}Chromium-release assays were performed with the MEFs. ^{51}Chromium-release assays are considered to show perforin-dependent lysis as the release of ^{51}Chromium just requires the disruption of the cell membrane. This disruption can be achieved by the formation of pores by perforin but also in the final stages of apoptosis induced by granzymes after about 2 hrs and 20 min of incubation with killer cells as determined by time-lapse microscopy (Waterhouse et al. 2006b).

The increased sulphation of the HSPG on the cell surface of cells double-deficient for *Sulf1* and *Sulf2* led to a slightly but significantly higher lysis by antigen-specific CTLs in comparison to Wt cells with a p-value of less than 0.001 as determined with ANOVA. Most of the phenotype of the *Sulf* Dko cell line could be accredited to the deficiency of *Sulf1*, as comparisons of Wt and *Sulf1*$^{-/-}$ cells showed (p < 0.01). Although, the uptake of GrB was not increased by elevated levels of sulphation of HS chains on MEFs deficient for *Sulf1* and *Sulf2*, the perforin-dependent lysis seemed to be increased especially in the double-deficient cell line. In all cytotoxic experiments with the cell lines deficient for *Sulf1* and *Sulf2* a control with EGTA was performed. EGTA is a chelating agent of divalent cations such as Ca^{2+} and can remove those cations so that granule secretion is blocked and no calcium-dependent killing can take place. The killing was always very low in this control meaning that hardly any killing not being calcium-dependent takes place and therefore the granule-exocytosis pathway is the pathway mainly used to lyse target cells. A second negative control was performed in which no SIINFEKL peptide was added during the 4 hrs of co-incubation of CTLs and target cells. The absence of killing in these controls indicated that no peptide-independent killing took place, meaning that only peptide-specific CTLs caused the lysis of target cells.

Next, the susceptibility of the *Sulf* clones towards apoptosis induced by peptide-specific CTLs was determined in [^{3}H]-Thymidine assays, which are considered to be GrB-dependent. [^{3}H]-Thymidine-release assays indicate apoptotic DNA fragmentation. No difference between the *Sulf* double-deficient and the Wt clones could be detected here (p = 0.167). Also for the cell lines deficient for either *Sulf1* (p = 0.10) or *Sulf2* (p = 0.15) no difference in killing could be

6.2 Role of sulphatases 1 and 2 in apoptosis

detected in comparison to Wt cells. In general, the percentage of apoptosis was comparibly low in those tests and even the stimulation of target cells with IFN-γ to increase H2K^b expression did not yield better results. The results of the [^{3}H]-Thymidine assays are therefore not fully conclusive. For all cytotoxic assays with *Sulf*-deficient cells, a Wt was chosen having a nearly identical H2K^b expression. Thus a difference in lysis between the cell lines caused by differential expression of the SIINFEKL restriction element H2K^b can be excluded.

Therefore, a specific GrB assay was performed, where GrB was delivered into the *Sulf* MEFs by using AdV type 5 as an endosomolytic agent to induce GrB-dependent apoptosis, which was analysed by sub G1-peak analysis in flow cytometry. No significant difference in GrB-induced apoptosis was detected between the 4 differentially sulphated MEFs clones.

Thus, no difference in apopototic killing was found in [^{3}H]-Thymidine release assays and in GrB-induced apoptosis between the *Sulf* clones. This is in accordance with the findings that the uptake of GrB did not differ between the single *Sulf* clones. The difference in lysis between the *Sulf* double-deficient and the *Sulf1*-deficient clone in comparison to the Wt in ^{51}Chromium-release assays is therefore assumed to be perforin-dependent. This is in accordance with literature. It was recently shown that CHO mutant cells A745 (lacking xylosyl transferase and therefore HSPGs) showed less membrane permeabilisation by perforin than Wt cells did, implying that perforin might bind to HS as well (Metkar et al. 2005). Thus, perforin might bind with different affinity to higher sulphated HSs and a higher sulphation leads to a higher lysis. *Sulf1*-deficiency led to increased sulphation on target cells and might cause increased perforin-dependent lysis. In contrast to granzymes, perforin is not redundant but instead essential for CTLs and NK cells as mentioned in the introduction.

Thus, an alteration of perforin binding by an alteration of HS sulphation patterns might be functionally important.

Generally, one can say that the question of how GrB is taken-up into cells is still not ultimately solved, whereby it is very likely that more than one pathway exists. For example, mutants of GrB reduced in their ability to bind to HS were not endocytosed but still successfully delivered into the cytosol of the target cells. Probably translocation into the cytosol of target cells is achieved by a process involving repairable membrane pores (Kurschus et al. 2008).

6.2.2 Uptake of adenovirus type 5 into mouse embryonic fibroblasts deficient for *Sulf1* and *Sulf2*

During the studies of GrB-induced apoptosis, perforin was substituted with AdV. Further analysis revealed that the productive infection of AdV-GFP was increased in *Sulf*-deficient cells.

Cell surface HSs are linear polysaccharides, which are strongly anionic but less sulphated

6 Discussion

than heparin (Jorpes and Gardell 1948). They can trap a wide variety of proteins and ligands, e.g. growth factors, coagulation factors, chemokines, cytokines but also pathogens (Bernfield et al. 1999). Among these pathogens are also viruses. HS mainly acts as co-receptor for viruses such as HIV-1 (Patel et al. 1993), gammaherpes virus 68 (Jarousse and Coscoy 2008) and other Herpes viruses as reviewed in (Shukla and Spear 2001), Dengue virus (Chen et al. 1997), Hepatitis B virus (Leistner et al. 2008), Foot and Mouth disease virus (Jackson et al. 1996) and Adenoviruses (Dechecchi et al. 2001, 2000).

We were able to show that the addition of AdV type 5 to *Sulf* MEFs resulted in elevated levels of GFP expression in *Sulf*-deficient in comparison to Wt MEFs. The GFP expression in the first set of *Sulf* MEFs was about 3.5-fold higher in the *Sulf* single-deficient clones and about 6.5-fold higher in the *Sulf* double deficient clone in comparison to the Wt clone at an MOI of 500. In the second set of independent *Sulf* MEFs, the GFP expression of the AdV was about 6.25-fold higher in the $Sulf2^{-/-}$ cell line, about 8-fold higher in the *Sulf* double-deficient cell line, and about 15-fold higher in the $Sulf1^{-/-}$ cell line in comparison to the Wt clone. The data indicate a dramatic increase in the productive infection of AdV type 5 in MEFs deficient for the *Sulf1* or *Sulf2* gene or both. This is quiet surprising as the infection of AdV is regarded as a two-step process involving distinct receptors in vitro. First, for attachment, AdV type 5 binds to a primary receptor, which was identified to be CAR (Bergelson et al. 1997; Tomko et al. 1997). Second, for the uptake of AdV type 5 the vitro-nectin binding integrins $\alpha_V\beta_3$ and $\alpha_V\beta_5$ are responsible (Wickham et al. 1993). Using antibodies against these integrins or antibodies blocking the penton base of the AdV required for the interaction with integrins inhibited the internalisation of the virus but not the binding (Wickham et al. 1993). Additionally, transfection of integrin α_V negative cells with cDNA encoding integrin α_V leads to its expression upon which AdV can successfully be internalised and infect the cells. These results indicate a strict separation between initial binding of the virus to the cell and its internalisation. Nevertheless, various other co-receptors were described as well (Hong et al. 1997; McDonald et al. 1999) and it is still a matter of discussion, whether they are involved in binding or uptake or not, e.g. MHC class I molecules.

The interaction of the AdV with the 46 kDa CAR is mediated via the fibre knob (indicated with B in figure 6.1 on page 132). To determine CAR expression in our cell system, cell lysates were analysed by immunoblots followed by densitometric analysis. In both sets of *Sulf* MEFs the *Sulf* double-deficient cells showed increased levels of CAR on their cell surface in comparison to the Wt clones as determined by densitometric analysis of immunoblots. The CAR expression was about 8-fold increased in the *Sulf* Dko in comparison to the Wt clone in the first set of MEFs, where the Dko showed the 6.5-fold higher GFP expression after addition of AdV. In the second set the increase in CAR expression between the *Sulf* double-deficient cell line and the Wt was about 1.6-fold.

6.2 Role of sulphatases 1 and 2 in apoptosis

For the lysates always exactly the same number of cells were seeded in plates to prevent differential expression of CAR due to differences in cell density, as CAR is known to be a transmembrane component of tight junctions. Sequestration of CAR in tight junctions inhibits infection of cells with AdV type 5 (Cohen et al. 2001), which is more common if cells are confluently grown. Nevertheless, the elevated CAR expression levels in the *Sulf* double-deficient cell lines do not completely explain the dramatically increased GFP expression. The increased expression of GFP in the *Sulf1* and *Sulf2*-deficient clones cannot be explained by their CAR levels as they were not elevated in comparison to the Wt clone.

After binding of the AdV to the primary receptor CAR, the uptake is then mediated via the co-receptors integrin $\alpha_v\beta_3$ and $\alpha_v\beta_5$ (Wickham et al. 1993). Their activation triggers also clathrin-mediated endocytosis involving the adaptor protein 2 and the large GTPase dynamin (Meier and Greber 2004; Nemerow and Stewart 1999). The interaction of AdV with the integrins is mediated via the aa sequence RGD in the penton base of the AdV (indicated with C in figure 6.1 on the next page).

We therefore determined the cell surface integrin α_v expression on the two sets of *Sulf* MEFs in flow cytometry with an antibody against this cell surface molecule also known as CD51. The expression levels of integrin α_v were nearly identical in both *Sulf* Wt clones. In the first set of *Sulf* clones the expression of integrin α_v was reduced in all *Sulf*-deficient clones in comparison to the Wt clone. In the second set of *Sulf* MEFs the integrin α_v expression of the *Sulf2*-deficient clone was identical with the Wt clone, whereas the expression in the *Sulf1*-deficient clone was increased and in the *Sulf* double-deficient was reduced in comparison to the Wt clone. These results uniformly show lower or equal levels of integrin α_v expression in the *Sulf2* and the double-deficient cell lines, contrasting the GFP expression results after successful infection with AdV-GFP.

Summarising the analysis of the cell surface receptors, we found increased GFP expression levels of the *Sulf*-deficient cell lines in comparison to three Wt clones. Most probably, these results are not based on increased levels of CAR on the cell surface, as this phenotype could just be found for the double-deficient cell line. The *Sulf* single-deficient clones showed the same levels of CAR expression as the Wt clone. The expression levels of integrin α_v were only increased in the *Sulf1*-deficient clone in the second set of *Sulf* MEFs in comparison to the Wt. All other clones showed reduced or equal integrin α_v expression levels. Thus, the different sulphation patterns of the HS chains on the *Sulf* MEFs might be responsible for the difference in GFP expression after successful infection with AdV.

The role of HS in functioning as receptor for AdV type 5 is controversial. Dechecchi and colleagues described that the AdV type 5 are able to bind to HSs via the fibre shaft and that this binding not only improves uptake into host cells but is also sufficient for infection with AdV type 2 into cells deficient for CAR (Dechecchi et al. 2001, 2000). The relevance of the KKTK

6 Discussion

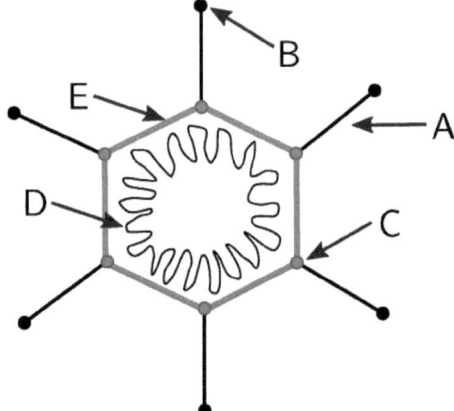

Figure 6.1: Schematic drawing of an AdV This cartoon illustrates an AdV, which belongs to the non-enveloped viruses with a diameter of 60–90 nm. **A** Fibre shaft with the recognition motif Lysine (K)KThreonine (T)K for HS; the fibre protein is 186 kDa in size **B** fibre knob with the recognition site for the primary receptor CAR **C** each 400 kDa pentavalent subunit (penton base) contains five Arginine (R)Gylcine (G)Aspartic acid (D) sequences, which are a binding motif for integrins **D** 36 kb linear DNA **E** Hexon protein. Information taken from (Modrow et al. 2003)

motif in the fibre shaft of the AdV (see A in figure 6.1) for binding to HS was investigated by several groups. The results are still conflicting as shaft mutations in a AdV type 5 precluded binding in vivo (Bayo-Puxan et al. 2006), whereas chimeric AdV type 5 with AdV type 31 and 41-derived fiber-shafts lacking the KKTK motif were as efficient in infection in vivo as the Wt AdV type 5 (Paolo et al. 2007). Chimeric adenoviral vectors, which do not use CAR for cell entrance also do not require HS, although the KKTK motif is present (Rogée et al. 2008). Rogée et al. (2008) propose chondroitin sulphate c as co-receptor.

To further assess the importance of HS in the infection process of AdV type 5 into MEFs, cells were treated with a mix of heparinases II and III to completely digest HS chains from syndecans and glypicans. We could demonstrate that treatment of MEFs with heparinases II and III did not decrease but in contrast improved the uptake of AdV type 5 into these cells (see figure 5.42 on page 115). The increase in GFP expression after treatment with heparinases was about 1.5-fold in all clones of the first *Sulf* MEF set and between 1.6 and 2.9-fold in the second set of *Sulf* MEFs. Thus, the removal of HS from the cell surface increases the uptake of AdV type 5 questioning the role of HSs as co-receptors in uptake.

A role for CAR in the uptake could even be excluded as determined in experiments with the CAR-deficient cell line CHO and the CHO mutant cell A745, which additionally lacks HSPGs. In these experiments AdV-GFP was added in different MOI to these two cell lines and the successful infection was analysed in flow cytometry as expression of GFP. The results

6.2 Role of sulphatases 1 and 2 in apoptosis

confirmed the data with the heparinase-treated *Sulf* MEFs, namely that the mutant CHO cell line A745, lacking HSPGs, showed a higher GFP expression after successful infection than the CHO cells (see figure 5.44 on page 116). Thus, the absence of HS leads to a better infection with AdV type 5 as determined by GFP expression.

We therefore propose an alternative model for HSs in the uptake of AdV type 5. HSs could act as "decoy receptors" for AdV type 5. Figure 6.2 on the next page illustrates how the decoy model could work. Part A of the figure depicts a cell with HSPGs and CAR on the cell surface and AdV ready for infecting the cell. Part B demonstrates the text book model, which works for other viruses besides AdV type 5, namely that HSs act as co-receptor for initial binding of virus particles to the cell and that these particles are then passed on to the primary receptor. In part C of figure 5.44 on page 116, the Dechecchi model describing the sufficiency of HS alone for the uptake of AdV is depicted. The alternative model we propose is illustrated in part D. Here, HSs on the cell surface snare virus particles and with this binding, AdV is trapped in the large side chains of the HSs, which prevents a binding of the AdV to its primary receptor CAR and therefore the uptake via the co-receptors integrin $\alpha_v\beta_3$ and $\alpha_v\beta_5$. So either HS just traps the AdV and prevents the binding to the actual receptor, or HS themselves are able to take up the AdV. These HS-bound AdV could then be fed into a non-infectious entry pathway as described for human papillomavirus (Selinka et al. 2007).

This decoy model just described explains our results, gained upon treatment of *Sulf* MEFs with heparinases and also to the results of the CHO clones with AdV infectivity, better than the existing model. If HS act as decoy receptors, they prevent the binding of AdV to CAR and α_v integrins, including translocation into an infectious entry pathway, which involves translation of viral genes, and thus, also of the transgenic GFP-fusion protein. After treating *Sulf* MEFs with heparinases II and III, which enzymatically digest most of the HS and heparin on the cell surface, the AdV particles can no longer be retained from CAR and integrin α_v and therefore we get a higher signal of GFP expression. In the case of the CHO mutant A745, a digestion of HS is not even required, but they show the same results: Absence of HS results in higher infection with AdV.

This new model must not be contradictory to the results of Dechecchi et al. (2001, 2000). They performed inhibition of infectivity experiments by using heparin as HS analog and found that heparin competes with cellular receptors for the binding of AdV. Furthermore, they incubated human alveolar type II-derived carcinoma A549 cells and HeLa cells with heparinases I–III and found that AdV type 5 uptake was reduced (Dechecchi et al. 2000). These results were gained by fluorescent foci count. For this approach, AdV was added to the cells. The cells were fixed and permeabilised and an antibody against a hexon protein of the AdV and a secondary antibody to visualise those proteins were used. Thus, this method shows the successful uptake of AdV into the cells but fails to indicate a productive infection. The other way of analysis,

6 Discussion

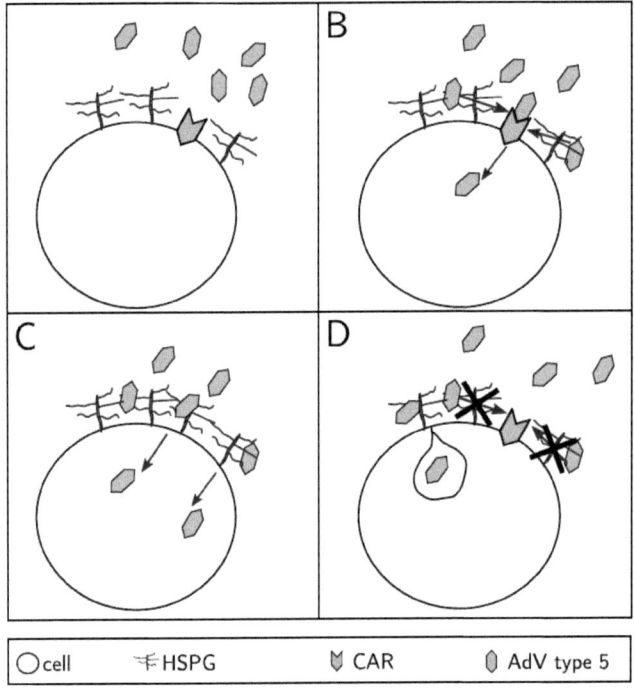

Figure 6.2: Models of uptake of AdV into cells This very simplified cartoon illustrates how AdV could be taken up into cells. A HS, as depicted by the small tree-like structures, can be found on all adherent cells and CAR, the light grey shapes, can also found on most cells. Upon an infection with AdV, the hexagons, distinct cell surface structures are involved in their uptake. **B** depicts the model described by Dechecchi (Dechecchi et al. 2000), where HSPGs act as co-receptors for AdV binding. **C** Dechecchi further describes that HSPGs are sufficient for initial binding and infection of AdV in CAR-deficient cells (Dechecchi et al. 2001). **D** We propose a new model, in which HSPGs act as decoy-receptors, meaning that they prevent the virus from binding to its actual receptor. Additionally, it could still be, that HSPGs mediate uptake of AdV itself, but into a non-infectious entry pathway leading to degradation of the virus.

which they used to study binding of the AdV to cells involved a radioactive pre-labelling of the AdV. Labelled AdV were added to the cells at 4 °C to just allow for binding and afterwards the cells were measured in a liquid scintillation counter. Also the second way of analysis besides the fluorescent foci count assay has the disadvantage, that it does not allow to distinguish between a non-infectious and an infectious entry pathway following binding.

We, on the other hand, used a genetically modified AdV encoding a GFP protein to investigate the successful uptake. The GFP gene requires the activation of the host machinery to be expressed and hence to be measured. No fluorescence signal from the GFP could be measured, if the AdV would be trapped in a non-infectious pathway.

6.2 Role of sulphatases 1 and 2 in apoptosis

The fact that HSs do not completely prevent the transfer of AdV type 5 to CAR and integrin α_V might be attributed to the differential locations of the binding sites for integrins (penton base), CAR (fibre knob), and potentially HSs (on the fibre shaft)(Smith et al. 2003; Wickham et al. 1993) (see figure 6.1 on page 132).

The existence of decoy receptors is a model, which is more accepted for chemokines and cytokines so far (Mantovani et al. 2001). Decoy receptors are defined as receptors, which can bind a ligand, but do not elicit a cellular response afterwards, so that they actually just prevent binding of the ligand to its high-affinity receptor. The most prominent example for a decoy receptor is the type II IL-1 receptor, which is not part of a signalling complex (Colotta et al. 1993). One new decoy receptor besides many accepted members of the TOLL/IL-1 R and TNFR families might be cell surface HSs. It remains to be analysed in how far the different sulphation patterns of HS play a critical role in the decoy receptor function.

The question which receptors are involved in the uptake and successful infection of AdV type 5 is of increasing importance as AdV are utilised for gene therapy, e.g. the treatment of malignant glioma (Immonen et al. 2004; Sandmair et al. 2000). One main restriction in the use of AdV for gene therapy is the rather low transfection efficiency of 5 to 68 % depending on tissue type, construct used, detection method, and study design (Laitinen et al. 1998; Nemunaitis et al. 2000; Puumalainen et al. 1998; Schuler et al. 2001). This can be caused by the use of the common serotypes 2 and 5 of group C, which are also the most common ones infecting humans (Bessis et al. 2004; Parks et al. 1999). Antibodies against wild type AdV from previous exposures can be found in 97 % of humans (Chirmule et al. 1999). Those antibodies together with the fact that AdV are highly immunogenic and activate the innate immune system, explain the low transfection efficiency. Another main restriction, which is linked to the first one is the liver tropism of human AdV type 5. More than 90 % of systemically delivered AdV is taken-up and destroyed by Kupffer cells (resident macrophages of the liver) in the liver (Alemany et al. 2000; Worgall et al. 1997). Although, AdV-vectors are underlying continuous improvement by removing more and more viral genes, adding tissue-specific promoters to activate other genes, and to detarget AdV from hepatocytes or from binding to their receptors CAR, integrins, and/or HSs ((Koizumi et al. 2006; Martin et al. 2003; Yun et al. 2005) and reviewed in (Dobbelstein 2004)), the design of AdV-vectors is still a big challenge. The findings that HSs might act as decoy receptors give the possibility to modify target cells in such a way, that the AdV-vector has better chances to be taken-up.

The receptors proposed to be involved in the binding and uptake of AdV type 5 or AdV-vectors based on this type or type 2 AdV vary depending on the type of cells or tissue analysed. CAR seems to be efficiently expressed on non-polarised cells in vitro but not apically on polarised cells, which more resembles the in vivo situation (Walters et al. 2002). In contrast, HSs are expressed apically on polarised epithelial cells (Mertens et al. 1996). Surprisingly, not

6 Discussion

only cell surface structures seem to play a role in the binding and uptake of AdV type 5 but also soluble blood factors such as the blood vitamin K-dependent coagulation factor IX, X, protein C, or the complement component C4-binding protein can bind to AdV (Parker et al. 2006). The soluble factors then bridge the binding of AdV to HSs and low density lipoprotein receptor-related proteins (Parker et al. 2006; Shayakhmetov et al. 2005). This interaction between the soluble factors and AdV type 5 is mediated by the hexon protein (indicated with an E in figure 6.1 on page 132) (Kalyuzhniy et al. 2008; Waddington et al. 2008). Not only these criteria but also the stage of cell cycle, in which the target cells are in at the time point of infection, seem to matter (Fender et al. 2008). Integrins are remodelled during mitosis, so that no internalisation of virus particles is possible at this stage of cell cycle.

The findings that a higher sulphation of the HSs due to the knock-out of *Sulf1* and *Sulf2* increased the uptake of AdV into *Sulf* MEFs point to the fact that the increased sulphation seen in the *Sulf*-deficient clones deteriorates the binding of AdV. It might be that *Sulf*-deficient cells have a reduced affinity for AdV type 5 or could just bind less AdV in comparison to *Sulf* Wt cells. In this scenario more AdV could be available for binding to integrin α_v or CAR. That would explain, why we see an increased production of GFP with the *Sulf*-deficient cells after addition of AdV for 24 hrs. Why the degradation of HSs on the cell surface by heparinases II and III equally rises the GFP production in all clones by a factor of about 1.5-fold in one set of MEFs and about 1.6 to 2.9-fold in the second set needs to be investigated further. However, complete removal of the HSs could further increase the chance of binding to receptors which mediate infection.

To solve the open question whether the deficiency of the *Sulf* genes reduces or increases the binding of AdV type 5 to HSs, shedded HSs from the Wt or *Sulf* single or double-deficient cells would need to be immobilised. Then AdV type 5 could be added and the exact amount of AdV binding to the HSs could be analysed e.g. with a biacore system.

To determine whether the increased uptake of AdV into cells expressing higher sulphated HSs also occurs in vivo, animal experiments were already started. In view of the fact that no interaction of AdV type 5 with CAR or integrin α_v is required for entry into hepatocytes in vivo (Alemany and Curiel 2001; Smith et al. 2002), these animal experiments might reveal new insights of the role of HSs in the infection process. Wt and *Sulf*-deficient mice were transfected with AdV expressing β-gal. Mice were sacrificed 72 hrs after transfection and the activity of β-gal was determined in preleminary experiments. With these experiments it will be possible to analyse in vivo, whether the different sulphation patterns have an influence on AdV infectivity in different tissue types.

Summary

Intracellular heat shock protein 70 (HSP70) belongs to the stress response system and can protect cells from different apoptotic stimuli by inhibiting apoptosis at different steps. Extracellular HSP70 on the other hand can act as an immunological danger signal and activate cells of the innate and adaptive immune system. Interestingly, intracellular HSP70 does not appear to protect cells against cytotoxic T-lymphocyte (CTL)-mediated killing. Acute overexpression of HSP70 could even increase the susceptibility of Ge-tet melanoma cells to CTLs that use the granule-exocytosis pathway for killing (Dressel et al. 1999).

To decipher the molecular pathway of this increased susceptibility to CTLs, the effect of the acute overexpression of HSP70 on gene expression was analysed in Ge-tet-1 cells, in which *Hsp70* is under the control of a tetracycline-inducible promotor. A gene expression profiling experiment using whole human genome arrays indicated that only 75 genes were significantly regulated upon HSP70 overexpression by addition of doxycycline. In a control clone (Ge-tra), containing only the transactivator domain, 43 genes were significantly regulated. Only 2 genes were significantly regulated in both clones by doxycycline. 13 genes were selected for confirmation by quantitative real-time PCR (qRT-PCR) experiments using RNA from parental Ge, Ge-tra, Ge-tet-1, and Ge-tet-2 cells. Only the up-regulation of *Hsp70* in the Ge-tet clones was confirmed both qualitatively and quantitatively. In summary, the acute overexpression of HSP70 does not appear to have a major impact on the regulation of other genes and the effects of HSP70 on CTL-mediated killing are regulated most likely on the protein level.

To reduce the complexity of CTL-induced apoptosis, further experiments were performed with granzyme (Gr)B, a component of the cytotoxic granules of CTLs and NK cells that has been shown to interact with HSP70. Acute overexpression of HSP70 significantly increased the percentage of Ge-tet-2 cells with fragmented DNA upon GrB-induced apoptosis as determined by sub-G1 peak analysis in flow cytometry. Furthermore, the acute overexpression of HSP70 slightly increased the percentage of Ge-tet-1 cells with a reduced mitochondrial cytochrome c content after GrB-induced apoptosis. These results suggest that HSP70 is not inhibiting GrB-induced apoptosis but can slightly enhance the progression of apoptosis in some cells. In staurosporine-induced apoptosis, in contrast, the acute overexpression of HSP70 partly protected Ge-tet-1 cells from early apoptosis. However, the small effects of HSP70 found in GrB-induced apoptosis do not sufficiently explain the increased susceptibility of these cells after

Summary

acute HSP70 overexpression against CTL-mediated killing.

Heparan sulphates (HSs) are involved in GrB binding to target cells of CTLs and Raja et al. (2005) could show that a suppression of sulphation of HSs on the cell surface led to a reduced uptake of GrB and diminished GrB and CTL-induced apoptosis. The sulphatases 1 and 2 modify the sulphation pattern of HSs. A deficiency of *Sulf1* and *Sulf2* genes could therefore have an effect on the efficiency of CTL-mediated apoptosis, as *Sulf*-deficient cells show increased sulphation patterns on HS proteoglycans (PGs).

Mouse embryonic fibroblasts (MEFs) from *Sulf1* and *Sulf2* double knock-out mice were used as targets for CTLs and their susceptibility to granule-exocytosis-dependent killing was slightly but significantly increased. This could be attributed to the deficiency of the *Sulf1* gene. GrB-induced apoptosis was not significantly altered suggesting no major contribution of *Sulf1* and *Sulf2* genes to the control of susceptibility of target cells to GrB and CTL-induced apoptosis.

However, during these experiments it turned out that the deficiency of *Sulf1* and *Sulf2* or both drastically increased the infection of these cells with adenovirus (AdV) type 5 that was used to deliver the GrB. Target cell infection was determined by green fluorescence protein (GFP) expression since the AdV contained a GFP expression construct. This finding was not explained by a different expression of receptors known to be involved in the uptake of AdV (coxsackie and adenovirus receptor (CAR) and integrin α_v) on the *Sulf*-deficient MEFs.

Since HSPGs had been suggested to function as co-receptors for AdV, the *Sulf*-deficiency might improve this co-receptor function. However, the digestion of HSPGs by heparinases II and III further increased the efficacy of infection. Similarly, a genetic defect of HSPGs in CHO cells was associated with an increased infection rate. These results suggest that HSPGs could function as decoy receptors for AdV channeling them into a non-infectious pathway rather than functioning as co-receptors for infection. Therefore, a modification of HSPGs on target cells might be a strategy to improve infection rates in AdV-based gene therapy.

Bibliography

Abravaya, K., Myers, M.P., Murphy, S.P. and Morimoto, R.I. (1992). The human heat shock protein hsp70 interacts with HSF, the transcription factor that regulates heat shock gene expression. *Genes Dev*, **6**, 1153–1164.

Adrain, C., Duriez, P.J., Brumatti, G., Delivani, P. and Martin, S.J. (2006). The cytotoxic lymphocyte protease, granzyme B, targets the cytoskeleton and perturbs microtubule polymerization dynamics. *J Biol Chem*, **281**, 8118–8125.

Ahn, J.H., Ko, Y.G., Park, W.Y., Kang, Y.S., Chung, H.Y. and Seo, J.S. (1999). Suppression of ceramide-mediated apoptosis by HSP70. *Mol Cells*, **9**, 200–6.

Ai, X., Do, A.T., Kusche-Gullberg, M., Lindahl, U., Lu, K. and Emerson, C. P., J. (2006). Substrate specificity and domain functions of extracellular heparan sulfate 6-O-endosulfatases, QSulf1 and QSulf2. *J Biol Chem*, **281**, 4969–76.

Ai, X., Do, A.T., Lozynska, O., Kusche-Gullberg, M., Lindahl, U. and Emerson, C. P., J. (2003). QSulf1 remodels the 6-O sulfation states of cell surface heparan sulfate proteoglycans to promote Wnt signaling. *J Cell Biol*, **162**, 341–51.

Alemany, R. and Curiel, D.T. (2001). CAR-binding ablation does not change biodistribution and toxicity of adenoviral vectors. *Gene Ther*, **8**, 1347–1353.

Alemany, R., Suzuki, K. and Curiel, D.T. (2000). Blood clearance rates of adenovirus type 5 in mice. *J Gen Virol*, **81**, 2605–2609.

Alimonti, J.B., Shi, L., Baijal, P.K. and Greenberg, A.H. (2001). Granzyme B induces BID-mediated cytochrome c release and mitochondrial permeability transition. *J Biol Chem*, **276**, 6974–82.

Andrade, F., Roy, S., Nicholson, D., Thornberry, N., Rosen, A. and Casciola-Rosen, L. (1998). Granzyme B directly and efficiently cleaves several downstream caspase substrates: implications for CTL-induced apoptosis. *Immunity*, **8**, 451–60.

Andrade, F., Fellows, E., Jenne, D.E., Rosen, A. and Young, C.S.H. (2007). Granzyme H destroys the function of critical adenoviral proteins required for viral DNA replication and granzyme B inhibition. *EMBO J*, **26**, 2148–2157.

Bibliography

Asea, A., Kraeft, S.K., Kurt-Jones, E.A., Stevenson, M.A., Chen, L.B., Finberg, R.W., Koo, G.C. and Calderwood, S.K. (2000). HSP70 stimulates cytokine production through a CD14-dependant pathway, demonstrating its dual role as a chaperone and cytokine. *Nat Med*, **6**, 435–42.

Atkinson, E.A., Barry, M., Darmon, A.J., Shostak, I., Turner, P.C., Moyer, R.W. and Bleackley, R.C. (1998). Cytotoxic T lymphocyte-assisted suicide. Caspase 3 activation is primarily the result of the direct action of granzyme B. *J Biol Chem*, **273**, 21261–6.

Barnhart, B.C., Alappat, E.C. and Peter, M.E. (2003). The CD95 type I/type II model. *Semin Immunol*, **15**, 185–193.

Barry, M. and Bleackley, R.C. (2002). Cytotoxic T lymphocytes: all roads lead to death. *Nat Rev Immunol*, **2**, 401–9.

Barry, M., Heibein, J.A., Pinkoski, M.J., Lee, S.F., Moyer, R.W., Green, D.R. and Bleackley, R.C. (2000). Granzyme B short-circuits the need for caspase 8 activity during granule-mediated cytotoxic T-lymphocyte killing by directly cleaving Bid. *Mol Cell Biol*, **20**, 3781–94.

Basu, S., Binder, R.J., Ramalingam, T. and Srivastava, P.K. (2001). CD91 is a common receptor for heat shock proteins gp96, hsp90, hsp70, and calreticulin. *Immunity*, **14**, 303–13.

Bayo-Puxan, N., Cascallo, M., Gros, A., Huch, M., Fillat, C. and Alemany, R. (2006). Role of the putative heparan sulfate glycosaminoglycan-binding site of the adenovirus type 5 fiber shaft on liver detargeting and knob-mediated retargeting. *J Gen Virol*, **87**, 2487–2495.

Becker, T.C., Noel, R.J., Coats, W.S., Gomez-Foix, A.M., Alam, T., Gerard, R.D. and Newgard, C.B. (1994). Use of recombinant adenovirus for metabolic engineering of mammalian cells. *Methods Cell Biol*, **43 Pt A**, 161–89.

Beere, H.M. (2004). "The stress of dying": the role of heat shock proteins in the regulation of apoptosis. *J Cell Sci*, **117**, 2641–51.

Beere, H.M. (2005). Death versus survival: functional interaction between the apoptotic and stress-inducible heat shock protein pathways. *J Clin Invest*, **115**, 2633–9.

Beere, H.M., Wolf, B.B., Cain, K., Mosser, D.D., Mahboubi, A., Kuwana, T., Tailor, P., Morimoto, R.I., Cohen, G.M. and Green, D.R. (2000). Heat-shock protein 70 inhibits apoptosis by preventing recruitment of procaspase-9 to the Apaf-1 apoptosome. *Nat Cell Biol*, **2**, 469–75.

Beresford, P.J., Xia, Z., Greenberg, A.H. and Lieberman, J. (1999). Granzyme A loading induces rapid cytolysis and a novel form of DNA damage independently of caspase activation. *Immunity*, **10**, 585–94.

Bibliography

Bergelson, J.M., Cunningham, J.A., Droguett, G., Kurt-Jones, E.A., Krithivas, A., Hong, J.S., Horwitz, M.S., Crowell, R.L. and Finberg, R.W. (1997). Isolation of a common receptor for Coxsackie B viruses and adenoviruses 2 and 5. *Science*, **275**, 1320–3.

Bernfield, M., Götte, M., Park, P.W., Reizes, O., Fitzgerald, M.L., Lincecum, J. and Zako, M. (1999). Functions of cell surface heparan sulfate proteoglycans. *Annu Rev Biochem*, **68**, 729–777.

Bessis, N., GarciaCozar, F.J. and Boissier, M.C. (2004). Immune responses to gene therapy vectors: influence on vector function and effector mechanisms. *Gene Ther*, **11 Suppl 1**, S10–S17.

Bienemann, A.S., Lee, Y.B., Howarth, J. and Uney, J.B. (2008). Hsp70 suppresses apoptosis in sympathetic neurones by preventing the activation of c-Jun. *J Neurochem*, **104**, 271–278.

Binder, R.J., Han, D.K. and Srivastava, P.K. (2000). CD91: a receptor for heat shock protein gp96. *Nat Immunol*, **1**, 151–5.

Bird, C.H., Sun, J., Ung, K., Karambalis, D., Whisstock, J.C., Trapani, J.A. and Bird, P.I. (2005). Cationic sites on granzyme B contribute to cytotoxicity by promoting its uptake into target cells. *Mol Cell Biol*, **25**, 7854–67.

Bivik, C., Rosdahl, I. and Ollinger, K. (2007). Hsp70 protects against UVB induced apoptosis by preventing release of cathepsins and cytochrome c in human melanocytes. *Carcinogenesis*, **28**, 537–544.

Bossi, G. and Griffiths, G.M. (1999). Degranulation plays an essential part in regulating cell surface expression of Fas ligand in T cells and natural killer cells. *Nat Med*, **5**, 90–6.

Bouchentouf, M., Benabdallah, B.F. and Tremblay, J.P. (2004). Myoblast survival enhancement and transplantation success improvement by heat-shock treatment in mdx mice. *Transplantation*, **77**, 1349–1356.

Bromley, S.K., Burack, W.R., Johnson, K.G., Somersalo, K., Sims, T.N., Sumen, C., Davis, M.M., Shaw, A.S., Allen, P.M. and Dustin, M.L. (2001). The immunological synapse. *Annu Rev Immunol*, **19**, 375–96.

Browne, K.A., Blink, E., Sutton, V.R., Froelich, C.J., Jans, D.A. and Trapani, J.A. (1999). Cytosolic delivery of granzyme B by bacterial toxins: evidence that endosomal disruption, in addition to transmembrane pore formation, is an important function of perforin. *Mol Cell Biol*, **19**, 8604–15.

Brunner, K.T., Mauel, J., Cerottini, J.C. and Chapuis, B. (1968). Quantitative assay of the lytic action of immune lymphoid cells on 51-Cr-labelled allogeneic target cells in vitro; inhibition by isoantibody and by drugs. *Immunology*, **14**, 181–96.

Bibliography

Castellino, F., Boucher, P.E., Eichelberg, K., Mayhew, M., Rothman, J.E., Houghton, A.N. and Germain, R.N. (2000). Receptor-mediated uptake of antigen/heat shock protein complexes results in major histocompatibility complex class I antigen presentation via two distinct processing pathways. *J Exp Med*, **191**, 1957–64.

Cellsignal (2008). Apoptosis overview. URL http://www.cellsignal.com/reference/pathway/ Apoptosis_Overview.html last checked 13rd February 2009.

Chae, H.J., Kang, J.S., Byun, J.O., Han, K.S., Kim, D.U., Oh, S.M., Kim, H.M., Chae, S.W. and Kim, H.R. (2000). Molecular mechanism of staurosporine-induced apoptosis in osteoblasts. *Pharmacol Res*, **42**, 373–381.

Chaitanya, G.V. and Babu, P.P. (2008). Multiple apoptogenic proteins are involved in the nuclear translocation of Apoptosis Inducing Factor during transient focal cerebral ischemia in rat. *Brain Res*, **1246**, 178–190.

Charlot, J.F., Prétet, J.L., Haughey, C. and Mougin, C. (2004). Mitochondrial translocation of p53 and mitochondrial membrane potential (Delta Psi m) dissipation are early events in staurosporine-induced apoptosis of wild type and mutated p53 epithelial cells. *Apoptosis*, **9**, 333–343.

Chen, C.J., Kono, H., Golenbock, D., Reed, G., Akira, S. and Rock, K.L. (2007). Identification of a key pathway required for the sterile inflammatory response triggered by dying cells. *Nat Med*, **13**, 851–856.

Chen, L., Willis, S.N., Wei, A., Smith, B.J., Fletcher, J.I., Hinds, M.G., Colman, P.M., Day, C.L., Adams, J.M. and Huang, D.C.S. (2005). Differential targeting of prosurvival Bcl-2 proteins by their BH3-only ligands allows complementary apoptotic function. *Mol Cell*, **17**, 393–403.

Chen, Y., Maguire, T., Hileman, R.E., Fromm, J.R., Esko, J.D., Linhardt, R.J. and Marks, R.M. (1997). Dengue virus infectivity depends on envelope protein binding to target cell heparan sulfate. *Nat Med*, **3**, 866–871.

Cheng, E.H., Kirsch, D.G., Clem, R.J., Ravi, R., Kastan, M.B., Bedi, A., Ueno, K. and Hardwick, J.M. (1997). Conversion of Bcl-2 to a Bax-like death effector by caspases. *Science*, **278**, 1966–1968.

Chirmule, N., Propert, K., Magosin, S., Qian, Y., Qian, R. and Wilson, J. (1999). Immune responses to adenovirus and adeno-associated virus in humans. *Gene Ther*, **6**, 1574–1583.

Ciocca, D.R., Fuqua, S.A., Lock-Lim, S., Toft, D.O., Welch, W.J. and McGuire, W.L. (1992). Response of human breast cancer cells to heat shock and chemotherapeutic drugs. *Cancer Res*, **52**, 3648–54.

Clark, P.R. and Menoret, A. (2001). The inducible Hsp70 as a marker of tumor immunogenicity. *Cell Stress Chaperones*, **6**, 121–5.

Bibliography

Clarke, P.G. (1990). Developmental cell death: morphological diversity and multiple mechanisms. *Anat Embryol (Berl)*, **181**, 195–213.

Cohen, C.J., Shieh, J.T., Pickles, R.J., Okegawa, T., Hsieh, J.T. and Bergelson, J.M. (2001). The coxsackievirus and adenovirus receptor is a transmembrane component of the tight junction. *Proc Natl Acad Sci U S A*, **98**, 15191–15196.

Cohen, J.J., Duke, R.C., Fadok, V.A. and Sellins, K.S. (1992). Apoptosis and programmed cell death in immunity. *Annu Rev Immunol*, **10**, 267–293.

Coleman, M.L., Sahai, E.A., Yeo, M., Bosch, M., Dewar, A. and Olson, M.F. (2001). Membrane blebbing during apoptosis results from caspase-mediated activation of ROCK I. *Nat Cell Biol*, **3**, 339–345.

Colotta, F., Re, F., Muzio, M., Bertini, R., Polentarutti, N., Sironi, M., Giri, J.G., Dower, S.K., Sims, J.E. and Mantovani, A. (1993). Interleukin-1 type II receptor: a decoy target for IL-1 that is regulated by IL-4. *Science*, **261**, 472–475.

Cossarizza, A., Baccarani-Contri, M., Kalashnikova, G. and Franceschi, C. (1993). A new method for the cytofluorimetric analysis of mitochondrial membrane potential using the J-aggregate forming lipophilic cation 5,5',6,6'-tetrachloro-1,1',3,3'-tetraethylbenzimidazolcarbocyanine iodide (JC-1). *Biochem Biophys Res Commun*, **197**, 40–45.

Cossarizza, A., Kalashnikova, G., Grassilli, E., Chiappelli, F., Salvioli, S., Capri, M., Barbieri, D., Troiano, L., Monti, D. and Franceschi, C. (1994). Mitochondrial modifications during rat thymocyte apoptosis: a study at the single cell level. *Exp Cell Res*, **214**, 323–330.

Cotter, T.G., Lennon, S.V., Glynn, J.M. and Green, D.R. (1992). Microfilament-disrupting agents prevent the formation of apoptotic bodies in tumor cells undergoing apoptosis. *Cancer Res*, **52**, 997–1005.

Darmon, A.J., Nicholson, D.W. and Bleackley, R.C. (1995). Activation of the apoptotic protease CPP32 by cytotoxic T-cell-derived granzyme B. *Nature*, **377**, 446–8.

Davis, D.M. (2002). Assembly of the immunological synapse for T cells and NK cells. *Trends Immunol*, **23**, 356–63.

Debatin, K.M., Stahnke, K. and Fulda, S. (2003). Apoptosis in hematological disorders. *Semin Cancer Biol*, **13**, 149–158.

Dechecchi, M.C., Melotti, P., Bonizzato, A., Santacatterina, M., Chilosi, M. and Cabrini, G. (2001). Heparan sulfate glycosaminoglycans are receptors sufficient to mediate the initial binding of adenovirus types 2 and 5. *J Virol*, **75**, 8772–8780.

Bibliography

Dechecchi, M.C., Tamanini, A., Bonizzato, A. and Cabrini, G. (2000). Heparan sulfate glycosaminoglycans are involved in adenovirus type 5 and 2-host cell interactions. *Virology*, **268**, 382–390.

Degli-Esposti, M.A. and Smyth, M.J. (2005). Close encounters of different kinds: dendritic cells and NK cells take centre stage. *Nat Rev Immunol*, **5**, 112–124.

Degterev, A. and Yuan, J. (2008). Expansion and evolution of cell death programmes. *Nat Rev Mol Cell Biol*, **9**, 378–390.

Dhoot, G.K., Gustafsson, M.K., Ai, X., Sun, W., Standiford, D.M. and Emerson, C.P. (2001). Regulation of Wnt signaling and embryo patterning by an extracellular sulfatase. *Science*, **293**, 1663–1666.

Dierker, T., Dreier, R., Petersen, A., Bordych, C. and Grobe, K. (2009). Heparan sulfate modulated, metalloprotease mediated sonic hedgehog release from producing cells. doi:10.1074/jbc.M806838200.

Djerassi, C. (1993). *Cantors Dilemma*. Heyne.

Dobbelstein, M. (2004). Replicating adenoviruses in cancer therapy. *Curr Top Microbiol Immunol*, **273**, 291–334.

Dressel, R., Elsner, L., Quentin, T., Walter, L. and Günther, E. (2000). Heat shock protein 70 is able to prevent heat shock-induced resistance of target cells to CTL. *J Immunol*, **164**, 2362–2371.

Dressel, R., Grzeszik, C., Kreiss, M., Lindemann, D., Herrmann, T., Walter, L. and Günther, E. (2003). Differential effect of acute and permanent heat shock protein 70 overexpression in tumor cells on lysability by cytotoxic T lymphocytes. *Cancer Res*, **63**, 8212–8220.

Dressel, R. and Günther, E. (1999). Heat-induced expression of MHC-linked HSP70 genes in lymphocytes varies at the single-cell level. *J Cell Biochem*, **72**, 558–569.

Dressel, R., Heine, L., Elsner, L., Geginat, G., Gefeller, O., Kölmel, K.F. and Günther, E. (1996). Induction of heat shock protein 70 genes in human lymphocytes during fever therapy. *Eur J Clin Invest*, **26**, 499–505.

Dressel, R., Johnson, J.P. and Günther, E. (1998). Heterogeneous patterns of constitutive and heat shock induced expression of HLA-linked HSP70-1 and HSP70-2 heat shock genes in human melanoma cell lines. *Melanoma Res*, **8**, 482–492.

Dressel, R., Lübbers, M., Walter, L., Herr, W. and Günther, E. (1999). Enhanced susceptibility to cytotoxic T lymphocytes without increase of MHC class I antigen expression after conditional overexpression of heat shock protein 70 in target cells. *Eur J Immunol*, **29**, 3925–3935.

Dressel, R., Raja, S.M., Höning, S., Seidler, T., Froelich, C.J., von Figura, K. and Günther, E. (2004a). Granzyme-mediated cytotoxicity does not involve the mannose 6-phosphate receptors on target cells. *J Biol Chem*, **279**, 20200–20210.

Bibliography

Dressel, R., von Figura, K. and Günther, E. (2004b). Unimpaired allorejection of cells deficient for the mannose 6-phosphate receptors Mpr300 and Mpr46. *Transplantation*, **78**, 758–761.

Dressel, R. and Demiroglu, S.Y. (2006). Role of Heat Shock Protein 70 in Apoptosis Mediated by Cytotoxic T Lymphocytes. In J. Radons and G. Multhoff, editors, *Heat Shock Proteins in Biology and Medicine*, pages 497–513. Research Signpost, Kerala. ISBN 81-308-0105-1.

Duan, H., Orth, K., Chinnaiyan, A.M., Poirier, G.G., Froelich, C.J., He, W.W. and Dixit, V.M. (1996). ICE-LAP6, a novel member of the ICE/Ced-3 gene family, is activated by the cytotoxic T cell protease granzyme B. *J Biol Chem*, **271**, 16720–16724.

Dudeja, V., Mujumdar, N., Phillips, P., Chugh, R., Borja-Cacho, D., Dawra, R.K., Vickers, S.M. and Saluja, A.K. (2009). Heat shock protein70 inhibits apoptosis in cancer cells through simultaneous and independent mechanisms. doi:10.1053/j.gastro.2009.01.070.

Dudoit, S., Yang, Y.H., Callow, M.J. and Speed, T.P. (2002). Statistical methods for identifying differentially expressed genes in replicated cDNA microarray experiments. *Stat Sinica*, **12**, 111–139.

Elsner, L., Muppala, V., Gehrmann, M., Lozano, J., Malzahn, D., Bickeböller, H., Brunner, E., Zientkowska, M., Herrmann, T., Walter, L., Alves, F., Multhoff, G. and Dressel, R. (2007). The heat shock protein HSP70 promotes mouse NK cell activity against tumors that express inducible NKG2D ligands. *J Immunol*, **179**, 5523–33.

Elsner, L., Flügge, P.F., Lozano, J., Muppala, V., Eiz-Vesper, B., Demiroglu, S.Y., Malzahn, D., Herrmann, T., Brunner, E., Bickeböller, H., Multhoff, G., Walter, L. and Dressel, R. (2009). The endogenous danger signals HSP70 and MICA cooperate in the activation of cytotoxic effector functions of NK cells. doi:10.1111/j.1582-4934.2008.00677.x.

Enari, M., Sakahira, H., Yokoyama, H., Okawa, K., Iwamatsu, A. and Nagata, S. (1998). A caspase-activated DNase that degrades DNA during apoptosis, and its inhibitor ICAD. *Nature*, **391**, 43–50.

Fan, Z., Beresford, P.J., Oh, D.Y., Zhang, D. and Lieberman, J. (2003). Tumor suppressor NM23-H1 is a granzyme A-activated DNase during CTL-mediated apoptosis, and the nucleosome assembly protein SET is its inhibitor. *Cell*, **112**, 659–72.

Fawell, S., Seery, J., Daikh, Y., Moore, C., Chen, L.L., Pepinsky, B. and Barsoum, J. (1994). Tat-mediated delivery of heterologous proteins into cells. *Proc Natl Acad Sci U S A*, **91**, 664–668.

Fender, P., Schoehn, G., Perron-Sierra, F., Tucker, G.C. and Lortat-Jacob, H. (2008). Adenovirus dodecahedron cell attachment and entry are mediated by heparan sulfate and integrins and vary along the cell cycle. *Virology*, **371**, 155–164.

Bibliography

Fernandes-Alnemri, T., Armstrong, R.C., Krebs, J., Srinivasula, S.M., Wang, L., Bullrich, F., Fritz, L.C., Trapani, J.A., Tomaselli, K.J., Litwack, G. and Alnemri, E.S. (1996). In vitro activation of CPP32 and Mch3 by Mch4, a novel human apoptotic cysteine protease containing two FADD-like domains. *Proc Natl Acad Sci U S A*, **93**, 7464–7469.

Froelich, C.J., Orth, K., Turbov, J., Seth, P., Gottlieb, R., Babior, B., Shah, G.M., Bleackley, R.C., Dixit, V.M. and Hanna, W. (1996a). New paradigm for lymphocyte granule-mediated cytotoxicity. Target cells bind and internalize granzyme B, but an endosomolytic agent is necessary for cytosolic delivery and subsequent apoptosis. *J Biol Chem*, **271**, 29073–9.

Froelich, C.J., Turbov, J. and Hanna, W. (1996b). Human perforin: rapid enrichment by immobilized metal affinity chromatography (IMAC) for whole cell cytotoxicity assays. *Biochem Biophys Res Commun*, **229**, 44–9.

Fukazawa, T., Walter, B. and Owen-Schaub, L.B. (2003). Adenoviral Bid overexpression induces caspase-dependent cleavage of truncated Bid and p53-independent apoptosis in human non-small cell lung cancers. *J Biol Chem*, **278**, 25428–25434.

Fulda, S. and Debatin, K.M. (2004). Targeting apoptosis pathways in cancer therapy. *Curr Cancer Drug Targets*, **4**, 569–576.

Gabai, V.L., Mabuchi, K., Mosser, D.D. and Sherman, M.Y. (2002). Hsp72 and stress kinase c-jun N-terminal kinase regulate the bid-dependent pathway in tumor necrosis factor-induced apoptosis. *Mol Cell Biol*, **22**, 3415–24.

Gabai, V.L., Meriin, A.B., Mosser, D.D., Caron, A.W., Rits, S., Shifrin, V.I. and Sherman, M.Y. (1997). Hsp70 prevents activation of stress kinases. A novel pathway of cellular thermotolerance. *J Biol Chem*, **272**, 18033–7.

Gallucci, S., Lolkema, M. and Matzinger, P. (1999). Natural adjuvants: endogenous activators of dendritic cells. *Nat Med*, **5**, 1249–1255.

Garner, R., Helgason, C.D., Atkinson, E.A., Pinkoski, M.J., Ostergaard, H.L., Sorensen, O., Fu, A., Lapchak, P.H., Rabinovitch, A., McElhaney, J.E. and et al. (1994). Characterization of a granule-independent lytic mechanism used by CTL hybridomas. *J Immunol*, **153**, 5413–21.

Garrido, C., Gurbuxani, S., Ravagnan, L. and Kroemer, G. (2001). Heat shock proteins: endogenous modulators of apoptotic cell death. *Biochem Biophys Res Commun*, **286**, 433–42.

Gastpar, R., Gross, C., Rossbacher, L., Ellwart, J., Riegger, J. and Multhoff, G. (2004). The cell surface-localized heat shock protein 70 epitope TKD induces migration and cytolytic activity selectively in human NK cells. *J Immunol*, **172**, 972–80.

Bibliography

Geginat, G., Heine, L. and Günther, E. (1993). Effect of heat shock on susceptibility of normal lymphoblasts and of a heat shock protein 70-defective tumour cell line to cytotoxic T lymphocytes in vitro. *Scand J Immunol*, **37**, 314–21.

Gething, M.J. and Sambrook, J. (1992). Protein folding in the cell. *Nature*, **355**, 33–45.

Goldstein, J.S., Chen, T., Gubina, E., Pastor, R.W. and Kozlowski, S. (2000). ICAM-1 enhances MHC-peptide activation of CD8(+) T cells without an organized immunological synapse. *Eur J Immunol*, **30**, 3266–70.

Goping, I.S., Sawchuk, T., Underhill, D.A. and Bleackley, R.C. (2006). Identification of alpha-tubulin as a granzyme B substrate during CTL-mediated apoptosis. *J Cell Sci*, **119**, 858–865.

Gossen, M., Freundlieb, S., Bender, G., Müller, G., Hillen, W. and Bujard, H. (1995). Transcriptional activation by tetracyclines in mammalian cells. *Science*, **268**, 1766–9.

Griffiths, G.M. and Isaaz, S. (1993). Granzymes A and B are targeted to the lytic granules of lymphocytes by the mannose-6-phosphate receptor. *J Cell Biol*, **120**, 885–96.

Gromkowski, S.H., Yagi, J. and Janeway, C. A., J. (1989). Elevated temperature regulates tumor necrosis factor-mediated immune killing. *Eur J Immunol*, **19**, 1709–14.

Gross, C., Hansch, D., Gastpar, R. and Multhoff, G. (2003a). Interaction of heat shock protein 70 peptide with NK cells involves the NK receptor CD94. *Biol Chem*, **384**, 267–79.

Gross, C., Koelch, W., DeMaio, A., Arispe, N. and Multhoff, G. (2003b). Cell surface-bound heat shock protein 70 (Hsp70) mediates perforin-independent apoptosis by specific binding and uptake of granzyme B. *J Biol Chem*, **278**, 41173–81.

Grossman, W.J., Revell, P.A., Lu, Z.H., Johnson, H., Bredemeyer, A.J. and Ley, T.J. (2003). The orphan granzymes of humans and mice. *Curr Opin Immunol*, **5**, 544–552.

Grujic, M., Braga, T., Lukinius, A., Eloranta, M.L., Knight, S.D., Pejler, G. and Abrink, M. (2005). Serglycin-deficient cytotoxic T lymphocytes display defective secretory granule maturation and granzyme B storage. *J Biol Chem*, **280**, 33411–33418.

Grujic, M., Christensen, J.P., Sørensen, M.R., Abrink, M., Pejler, G. and Thomsen, A.R. (2008). Delayed contraction of the CD8+ T cell response toward lymphocytic choriomeningitis virus infection in mice lacking serglycin. *J Immunol*, **181**, 1043–1051.

Günther, E. and Walter, L. (1994). Genetic aspects of the hsp70 multigene family in vertebrates. *Experientia*, **50**, 987–1001.

Gurbuxani, S., Schmitt, E., Cande, C., Parcellier, A., Hammann, A., Daugas, E., Kouranti, I., Spahr, C., Pance, A., Kroemer, G. and Garrido, C. (2003). Heat shock protein 70 binding inhibits the nuclear import of apoptosis-inducing factor. *Oncogene*, **22**, 6669–78.

Bibliography

Habuchi, H., Tanaka, M., Habuchi, O., Yoshida, K., Suzuki, H., Ban, K. and Kimata, K. (2000). The occurrence of three isoforms of heparan sulfate 6-O-sulfotransferase having different specificities for hexuronic acid adjacent to the targeted N-sulfoglucosamine. *J Biol Chem*, **275**, 2859–2868.

Hada, H., Honda, C. and Tanemura, H. (1977). Spectroscopic study on the J-aggregate formation of cyanine dyes with molecule structures. *Photogr Sci Eng*, **21**, 83–91.

Hartl, F.U. and Hayer-Hartl, M. (2002). Molecular chaperones in the cytosol: from nascent chain to folded protein. *Science*, **295**, 1852–8.

Heibein, J.A., Goping, I.S., Barry, M., Pinkoski, M.J., Shore, G.C., Green, D.R. and Bleackley, R.C. (2000). Granzyme B-mediated cytochrome c release is regulated by the Bcl-2 family members bid and Bax. *J Exp Med*, **192**, 1391–402.

Henkart, P.A. (1985). Mechanism of lymphocyte-mediated cytotoxicity. *Annu Rev Immunol*, **3**, 31–58.

Henkart, P.A., Williams, M.S., Zacharchuk, C.M. and Sarin, A. (1997). Do CTL kill target cells by inducing apoptosis? *Semin Immunol*, **9**, 135–44.

Heusel, J.W., Wesselschmidt, R.L., Shresta, S., Russell, J.H. and Ley, T.J. (1994). Cytotoxic lymphocytes require granzyme B for the rapid induction of DNA fragmentation and apoptosis in allogeneic target cells. *Cell*, **76**, 977–87.

Hoehn, B., Ringer, T.M., Xu, L., Giffard, R.G., Sapolsky, R.M., Steinberg, G.K. and Yenari, M.A. (2001). Overexpression of HSP72 after induction of experimental stroke protects neurons from ischemic damage. *J Cereb Blood Flow Metab*, **21**, 1303–9.

Hogquist, K.A., Jameson, S.C., Heath, W.R., Howard, J.L., Bevan, M.J. and Carbone, F.R. (1994). T cell receptor antagonist peptides induce positive selection. *Cell*, **76**, 17–27.

Hong, S.S., Karayan, L., Tournier, J., Curiel, D.T. and Boulanger, P.A. (1997). Adenovirus type 5 fiber knob binds to MHC class I alpha2 domain at the surface of human epithelial and B lymphoblastoid cells. *EMBO J*, **16**, 2294–2306.

Honig, M.G. and Hume, R.I. (1986). Fluorescent carbocyanine dyes allow living neurons of identified origin to be studied in long-term cultures. *J Cell Biol*, **103**, 171–187.

Houge, G., Døskeland, S.O., Bøe, R. and Lanotte, M. (1993). Selective cleavage of 28S rRNA variable regions V3 and V13 in myeloid leukemia cell apoptosis. *FEBS Lett*, **315**, 16–20.

Hsieh, Y.C., Chang, M.S., Chen, J.Y., Yen, J.J.Y., Lu, I.C., Chou, C.M. and Huang, C.J. (2003). Cloning of zebrafish BAD, a BH3-only proapoptotic protein, whose overexpression leads to apoptosis in COS-1 cells and zebrafish embryos. *Biochem Biophys Res Commun*, **304**, 667–675.

Huang, D.C. and Strasser, A. (2000). BH3-Only proteins-essential initiators of apoptotic cell death. *Cell*, **103**, 839–842.

Bibliography

Huppa, J.B. and Davis, M.M. (2003). T-cell-antigen recognition and the immunological synapse. *Nat Rev Immunol*, **3**, 973–83.

Immonen, A., Vapalahti, M., Tyynelä, K., Hurskainen, H., Sandmair, A., Vanninen, R., Langford, G., Murray, N. and Ylä-Herttuala, S. (2004). AdvHSV-tk gene therapy with intravenous ganciclovir improves survival in human malignant glioma: a randomised, controlled study. *Mol Ther*, **10**, 967–972.

Jäättelä, M. (1999). Escaping cell death: survival proteins in cancer. *Exp Cell Res*, **248**, 30–43.

Jäättelä, M., Saksela, K. and Saksela, E. (1989). Heat shock protects WEHI-164 target cells from the cytolysis by tumor necrosis factors alpha and beta. *Eur J Immunol*, **19**, 1413–7.

Jäättelä, M. and Wissing, D. (1993). Heat-shock proteins protect cells from monocyte cytotoxicity: possible mechanism of self-protection. *J Exp Med*, **177**, 231–6.

Jäättelä, M., Wissing, D., Bauer, P.A. and Li, G.C. (1992). Major heat shock protein hsp70 protects tumor cells from tumor necrosis factor cytotoxicity. *EMBO J*, **11**, 3507–12.

Jäättelä, M., Wissing, D., Kokholm, K., Kallunki, T. and Egeblad, M. (1998). Hsp70 exerts its anti-apoptotic function downstream of caspase-3-like proteases. *EMBO J*, **17**, 6124–34.

Jackson, T., Ellard, F.M., Ghazaleh, R.A., Brookes, S.M., Blakemore, W.E., Corteyn, A.H., Stuart, D.I., Newman, J.W. and King, A.M. (1996). Efficient infection of cells in culture by type O foot-and-mouth disease virus requires binding to cell surface heparan sulfate. *J Virol*, **70**, 5282–5287.

Jarousse, N. and Coscoy, L. (2008). Selection of mutant CHO clones resistant to murine gammaherpesvirus 68 infection. *Virology*, **373**, 376–386.

Johnson, H., Scorrano, L., Korsmeyer, S.J. and Ley, T.J. (2003). Cell death induced by granzyme C. *Blood*, **101**, 3093–101.

Jolly, C. and Morimoto, R.I. (2000). Role of the heat shock response and molecular chaperones in oncogenesis and cell death. *J Natl Cancer Inst*, **92**, 1564–72.

Jorpes, J.E. and Gardell, S. (1948). On heparin monosulfuric acid. *J Biol Chem*, **176**, 267–276.

Kagi, D., Ledermann, B., Burki, K., Seiler, P., Odermatt, B., Olsen, K.J., Podack, E.R., Zinkernagel, R.M. and Hengartner, H. (1994). Cytotoxicity mediated by T cells and natural killer cells is greatly impaired in perforin-deficient mice. *Nature*, **369**, 31–7.

Kalyuzhniy, O., Paolo, N.C.D., Silvestry, M., Hofherr, S.E., Barry, M.A., Stewart, P.L. and Shayakhmetov, D.M. (2008). Adenovirus serotype 5 hexon is critical for virus infection of hepatocytes in vivo. *Proc Natl Acad Sci U S A*, **105**, 5483–5488.

Bibliography

Kerr, J.F., Wyllie, A.H. and Currie, A.R. (1972). Apoptosis: a basic biological phenomenon with wide-ranging implications in tissue kinetics. *Br J Cancer*, **26**, 239–257.

King, M.A., Eddaoudi, A. and Davies, D.C. (2007). A comparison of three flow cytometry methods for evaluating mitochondrial damage during staurosporine-induced apoptosis in Jurkat cells. *Cytometry A*, **71**, 668–674.

Koizumi, N., Kawabata, K., Sakurai, F., Watanabe, Y., Hayakawa, T. and Mizuguchi, H. (2006). Modified adenoviral vectors ablated for coxsackievirus-adenovirus receptor, alphav integrin, and heparan sulfate binding reduce in vivo tissue transduction and toxicity. *Hum Gene Ther*, **17**, 264–279.

Kuhn, J.R. and Poenie, M. (2002). Dynamic polarization of the microtubule cytoskeleton during CTL-mediated killing. *Immunity*, **16**, 111–21.

Kuo, C.C., Liang, S.M. and Liang, C.M. (2006). CpG-B oligodeoxynucleotide promotes cell survival via up-regulation of Hsp70 to increase Bcl-xL and to decrease apoptosis-inducing factor translocation. *J Biol Chem*, **281**, 38200–38207.

Kurschus, F.C., Bruno, R., Fellows, E., Falk, C.S. and Jenne, D.E. (2005). Membrane receptors are not required to deliver granzyme B during killer cell attack. *Blood*, **105**, 2049–58.

Kurschus, F.C., Fellows, E., Stegmann, E. and Jenne, D.E. (2008). Granzyme B delivery via perforin is restricted by size, but not by heparan sulfate-dependent endocytosis. *Proc Natl Acad Sci U S A*, **105**, 13799–804.

Kusher, D.I., Ware, C.F. and Gooding, L.R. (1990). Induction of the heat shock response protects cells from lysis by tumor necrosis factor. *J Immunol*, **145**, 2925–31.

Kuwana, T., Bouchier-Hayes, L., Chipuk, J.E., Bonzon, C., Sullivan, B.A., Green, D.R. and Newmeyer, D.D. (2005). BH3 domains of BH3-only proteins differentially regulate Bax-mediated mitochondrial membrane permeabilization both directly and indirectly. *Mol Cell*, **17**, 525–535.

Laemmli, U.K. (1970). Cleavage of structural proteins during the assembly of the head of bacteriophage T4. *Nature*, **227**, 680–685.

Laitinen, M., Mäkinen, K., Manninen, H., Matsi, P., Kossila, M., Agrawal, R.S., Pakkanen, T., Luoma, J.S., Viita, H., Hartikainen, J., Alhava, E., Laakso, M. and Ylä-Herttuala, S. (1998). Adenovirus-mediated gene transfer to lower limb artery of patients with chronic critical leg ischemia. *Hum Gene Ther*, **9**, 1481–1486.

Lamanna, W.C., Baldwin, R.J., Padva, M., Kalus, I., Ten Dam, G., van Kuppevelt, T.H., Gallagher, J.T., von Figura, K., Dierks, T. and Merry, C.L. (2006). Heparan sulfate 6-O-endosulfatases: discrete in vivo activities and functional co-operativity. *Biochem J*, **400**, 63–73.

Bibliography

Lamanna, W.C., Frese, M.A., Balleininger, M. and Dierks, T. (2008). Sulf loss influences N-, 2O-, and 6O-sulfation of multiple heparan sulfate proteoglycans and modulates FGF signaling. *J Biol Chem*, **283**, 27724–27735.

Lechardeur, D., Drzymala, L., Sharma, M., Zylka, D., Kinach, R., Pacia, J., Hicks, C., Usmani, N., Rommens, J.M. and Lukacs, G.L. (2000). Determinants of the nuclear localization of the heterodimeric DNA fragmentation factor (ICAD/CAD). *J Cell Biol*, **150**, 321–334.

Leistner, C.M., Gruen-Bernhard, S. and Glebe, D. (2008). Role of glycosaminoglycans for binding and infection of hepatitis B virus. *Cell Microbiol*, **10**, 122–133.

Letai, A., Bassik, M.C., Walensky, L.D., Sorcinelli, M.D., Weiler, S. and Korsmeyer, S.J. (2002). Distinct BH3 domains either sensitize or activate mitochondrial apoptosis, serving as prototype cancer therapeutics. *Cancer Cell*, **2**, 183–192.

Li, G.C. and Werb, Z. (1982). Correlation between synthesis of heat shock proteins and development of thermotolerance in Chinese hamster fibroblasts. *Proc Natl Acad Sci U S A*, **79**, 3218–22.

Li, H., Zhu, H., Xu, C.J. and Yuan, J. (1998). Cleavage of BID by caspase 8 mediates the mitochondrial damage in the Fas pathway of apoptosis. *Cell*, **94**, 491–501.

Li, L.Y., Luo, X. and Wang, X. (2001). Endonuclease G is an apoptotic DNase when released from mitochondria. *Nature*, **412**, 95–9.

Lieberman, J. (2003). The ABCs of granule-mediated cytotoxicity: new weapons in the arsenal. *Nat Rev Immunol*, **3**, 361–70.

Lin, X., Wei, G., Shi, Z., Dryer, L., Esko, J.D., Wells, D.E. and Matzuk, M.M. (2000). Disruption of gastrulation and heparan sulfate biosynthesis in EXT1-deficient mice. *Dev Biol*, **224**, 299–311.

Liu, L., Chen, J., Zhang, J., Ji, C., Zhang, X., Gu, S., Xie, Y. and Mao, Y. (2007). Overexpression of BimSs3, the novel isoform of Bim, can trigger cell apoptosis by inducing cytochrome c release from mitochondria. *Acta Biochim Pol*, **54**, 603–610.

Liu, Q.L., Kishi, H., Ohtsuka, K. and Muraguchi, A. (2003). Heat shock protein 70 binds caspase-activated DNase and enhances its activity in TCR-stimulated T cells. *Blood*, **102**, 1788–96.

Lohse, D.L. and Linhardt, R.J. (1992). Purification and characterization of heparin lyases from Flavobacterium heparinum. *J Biol Chem*, **267**, 24347–24355.

Lord, S.J., Rajotte, R.V., Korbutt, G.S. and Bleackley, R.C. (2003). Granzyme B: a natural born killer. *Immunol Rev*, **193**, 31–8.

Lowin, B., Beermann, F., Schmidt, A. and Tschopp, J. (1994a). A null mutation in the perforin gene impairs cytolytic T lymphocyte- and natural killer cell-mediated cytotoxicity. *Proc Natl Acad Sci U S A*, **91**, 11571–5.

Bibliography

Lowin, B., Hahne, M., Mattmann, C. and Tschopp, J. (1994b). Cytolytic T-cell cytotoxicity is mediated through perforin and Fas lytic pathways. *Nature*, **370**, 650–2.

López, E. and Ferrer, I. (2000). Staurosporine- and H-7-induced cell death in SH-SY5Y neuroblastoma cells is associated with caspase-2 and caspase-3 activation, but not with activation of the FAS/FAS-L-caspase-8 signaling pathway. *Brain Res Mol Brain Res*, **85**, 61–67.

Lüthi, A.U. and Martin, S.J. (2007). The CASBAH: a searchable database of caspase substrates. *Cell Death Differ*, **14**, 641–650.

MacDonald, G., Shi, L., Velde, C.V., Lieberman, J. and Greenberg, A.H. (1999). Mitochondria-dependent and -independent regulation of Granzyme B-induced apoptosis. *J Exp Med*, **189**, 131–144.

Mantovani, A., Locati, M., Vecchi, A., Sozzani, S. and Allavena, P. (2001). Decoy receptors: a strategy to regulate inflammatory cytokines and chemokines. *Trends Immunol*, **22**, 328–336.

Martin, K., Brie, A., Saulnier, P., Perricaudet, M., Yeh, P. and Vigne, E. (2003). Simultaneous CAR- and alpha V integrin-binding ablation fails to reduce Ad5 liver tropism. *Mol Ther*, **8**, 485–494.

Martin, S.J., Amarante-Mendes, G.P., Shi, L., Chuang, T.H., Casiano, C.A., O'Brien, G.A., Fitzgerald, P., Tan, E.M., Bokoch, G.M., Greenberg, A.H. and Green, D.R. (1996). The cytotoxic cell protease granzyme B initiates apoptosis in a cell-free system by proteolytic processing and activation of the ICE/CED-3 family protease, CPP32, via a novel two-step mechanism. *EMBO J*, **15**, 2407–2416.

Martinvalet, D., Zhu, P. and Lieberman, J. (2005). Granzyme A induces caspase-independent mitochondrial damage, a required first step for apoptosis. *Immunity*, **22**, 355–370.

Masson, D., Zamai, M. and Tschopp, J. (1986). Identification of granzyme A isolated from cytotoxic T-lymphocyte-granules as one of the proteases encoded by CTL-specific genes. *FEBS Lett*, **208**, 84–88.

Mayer, M.P., Brehmer, D., Gassler, C.S. and Bukau, B. (2001). Hsp70 chaperone machines. *Adv Protein Chem*, **59**, 1–44.

McConkey, D.J. (1998). Biochemical determinants of apoptosis and necrosis. *Toxicol Lett*, **99**, 157–168.

McDonald, D., Stockwin, L., Matzow, T., Blair Zajdel, M.E. and Blair, G.E. (1999). Coxsackie and adenovirus receptor (CAR)-dependent and major histocompatibility complex (MHC) class I-independent uptake of recombinant adenoviruses into human tumour cells. *Gene Ther*, **6**, 1512–9.

Bibliography

Medema, J.P., Toes, R.E., Scaffidi, C., Zheng, T.S., Flavell, R.A., Melief, C.J., Peter, M.E., Offringa, R. and Krammer, P.H. (1997). Cleavage of FLICE (caspase-8) by granzyme B during cytotoxic T lymphocyte-induced apoptosis. *Eur J Immunol*, **27**, 3492–8.

Mehlen, P., Schulze-Osthoff, K. and Arrigo, A.P. (1996). Small stress proteins as novel regulators of apoptosis. Heat shock protein 27 blocks Fas/APO-1- and staurosporine-induced cell death. *J Biol Chem*, **271**, 16510–16514.

Meier, O. and Greber, U.F. (2004). Adenovirus endocytosis. *J Gene Med*, **6 Suppl 1**, S152–S163.

Melcher, A., Todryk, S., Hardwick, N., Ford, M., Jacobson, M. and Vile, R.G. (1998). Tumor immunogenicity is determined by the mechanism of cell death via induction of heat shock protein expression. *Nat Med*, **4**, 581–7.

Menoret, A., Patry, Y., Burg, C. and Le Pendu, J. (1995). Co-segregation of tumor immunogenicity with expression of inducible but not constitutive hsp70 in rat colon carcinomas. *J Immunol*, **155**, 740–7.

Mertens, G., der Schueren, B.V., van den Berghe, H. and David, G. (1996). Heparan sulfate expression in polarized epithelial cells: the apical sorting of glypican (GPI-anchored proteoglycan) is inversely related to its heparan sulfate content. *J Cell Biol*, **132**, 487–497.

Metkar, S.S., Wang, B., Aguilar-Santelises, M., Raja, S.M., Uhlin-Hansen, L., Podack, E., Trapani, J.A. and Froelich, C.J. (2002). Cytotoxic cell granule-mediated apoptosis: perforin delivers granzyme B-serglycin complexes into target cells without plasma membrane pore formation. *Immunity*, **16**, 417–428.

Metkar, S.S., Wang, B., Ebbs, M.L., Kim, J.H., Lee, Y.J., Raja, S.M. and Froelich, C.J. (2003). Granzyme B activates procaspase-3 which signals a mitochondrial amplification loop for maximal apoptosis. *J Cell Biol*, **160**, 875–885.

Metkar, S.S., Wang, B. and Froelich, C.J. (2005). Detection of functional cell surface perforin by flow cytometry. *J Immunol Methods*, **299**, 117–27.

Modrow, S., Falke, D. and Truyen, U. (2003). *Molekulare Virologie*. Spektrum Akademischer Verlag.

Morimoto-Tomita, M., Uchimura, K., Werb, Z., Hemmerich, S. and Rosen, S.D. (2002). Cloning and characterization of two extracellular heparin-degrading endosulfatases in mice and humans. *J Biol Chem*, **277**, 49175–49185.

Mosser, D.D., Caron, A.W., Bourget, L., Meriin, A.B., Sherman, M.Y., Morimoto, R.I. and Massie, B. (2000). The chaperone function of hsp70 is required for protection against stress-induced apoptosis. *Mol Cell Biol*, **20**, 7146–59.

Bibliography

Mosser, D.D. and Morimoto, R.I. (2004). Molecular chaperones and the stress of oncogenesis. *Oncogene*, **23**, 2907–18.

Motyka, B., Korbutt, G., Pinkoski, M.J., Heibein, J.A., Caputo, A., Hobman, M., Barry, M., Shostak, I., Sawchuk, T., Holmes, C.F., Gauldie, J. and Bleackley, R.C. (2000). Mannose 6-phosphate/insulin-like growth factor II receptor is a death receptor for granzyme B during cytotoxic T cell-induced apoptosis. *Cell*, **103**, 491–500.

Mueller, O., Lightfoot, S. and Schroeder, A. (2004). RNA Integrity Number (RIN) – Standardization of RNA Quality Control. Agilent Technologies.

Mülhardt, C. (2006). *Der Experimentator: Molekularbiologie/Genomics*. Spektrum Akademischer Verlag, 5 edition.

Müllbacher, A., Ebnet, K., Blanden, R.V., Hla, R.T., Stehle, T., Museteanu, C. and Simon, M.M. (1996). Granzyme A is critical for recovery of mice from infection with the natural cytopathic viral pathogen, ectromelia. *Proc Natl Acad Sci U S A*, **93**, 5783–7.

Müllbacher, A., Waring, P., Tha Hla, R., Tran, T., Chin, S., Stehle, T., Museteanu, C. and Simon, M.M. (1999). Granzymes are the essential downstream effector molecules for the control of primary virus infections by cytolytic leukocytes. *Proc Natl Acad Sci U S A*, **96**, 13950–5.

Multhoff, G. (2002). Activation of natural killer cells by heat shock protein 70. *Int J Hyperthermia*, **18**, 576–85.

Multhoff, G., Botzler, C., Jennen, L., Schmidt, J., Ellwart, J. and Issels, R. (1997). Heat shock protein 72 on tumor cells: a recognition structure for natural killer cells. *J Immunol*, **158**, 4341–50.

Multhoff, G., Botzler, C., Wiesnet, M., Eissner, G. and Issels, R. (1995). CD3- large granular lymphocytes recognize a heat-inducible immunogenic determinant associated with the 72-kD heat shock protein on human sarcoma cells. *Blood*, **86**, 1374–82.

Multhoff, G., Mizzen, L., Winchester, C.C., Milner, C.M., Wenk, S., Eissner, G., Kampinga, H.H., Laumbacher, B. and Johnson, J. (1999). Heat shock protein 70 (Hsp70) stimulates proliferation and cytolytic activity of natural killer cells. *Exp Hematol*, **27**, 1627–36.

Multhoff, G., Pfister, K., Gehrmann, M., Hantschel, M., Gross, C., Hafner, M. and Hiddemann, W. (2001). A 14-mer Hsp70 peptide stimulates natural killer (NK) cell activity. *Cell Stress Chaperones*, **6**, 337–44.

Nagel, F., Dohm, C.P., Bähr, M., Wouters, F.S. and Dietz, G.P.H. (2008). Quantitative evaluation of chaperone activity and neuroprotection by different preparations of a cell-penetrating Hsp70. *J Neurosci Methods*, **171**, 226–232.

Bibliography

NCBI (2008). Basic local alignment search tool. URL http://blast.ncbi.nlm.nih.gov/Blast. cgi?PAGE=Nucleotides&PROGRAM=blastn&MEGABLAST=on&BLAST_PROGRAMS=megaBlast&PAGE_TYPE=BlastSearch&SHOW_DEFAULTS=on last checked 11th July 2008.

Neimanis, S., Albig, W., Doenecke, D. and Kahle, J. (2007). Sequence elements in both subunits of the DNA fragmentation factor are essential for its nuclear transport. *J Biol Chem*, **282**, 35821–35830.

Nemerow, G.R. and Stewart, P.L. (1999). Role of alpha(v) integrins in adenovirus cell entry and gene delivery. *Microbiol Mol Biol Rev*, **63**, 725–734.

Nemunaitis, J., Swisher, S.G., Timmons, T., Connors, D., Mack, M., Doerksen, L., Weill, D., Wait, J., Lawrence, D.D., Kemp, B.L., Fossella, F., Glisson, B.S., Hong, W.K., Khuri, F.R., Kurie, J.M., Lee, J.J., Lee, J.S., Nguyen, D.M., Nesbitt, J.C., Perez-Soler, R., Pisters, K.M., Putnam, J.B., Richli, W.R., Shin, D.M., Walsh, G.L., Merritt, J. and Roth, J. (2000). Adenovirus-mediated p53 gene transfer in sequence with cisplatin to tumors of patients with non-small-cell lung cancer. *J Clin Oncol*, **18**, 609–622.

Nicolier, M., Decrion-Barthod, A.Z., Launay, S., Prétet, J.L. and Mougin, C. (2009). Spatiotemporal activation of caspase-dependent and -independent pathways in staurosporine-induced apoptosis of p53wt and p53mt human cervical carcinoma cells. doi:10.1042/BC20080164.

Noble, P.B. and Cutts, J.H. (1967). Separation of blood leukocytes by Ficoll gradient. *Can Vet J*, **8**, 110–111.

Novota, P., Sviland, L., Zinöcker, S., Stocki, P., Balavarca, Y., Bickeböller, H., Rolstad, B., Wang, X.N., Dickinson, A.M. and Dressel, R. (2008). Correlation of Hsp70-1 and Hsp70-2 gene expression with the degree of graft-versus-host reaction in a rat skin explant model. *Transplantation*, **85**, 1809–1816.

Nyberg-Hoffman, C. and Aguilar-Cordova, E. (1999). Instability of adenoviral vectors during transport and its implication for clinical studies. *Nat Med*, **5**, 955–7.

Omura, S., Iwai, Y., Hirano, A., Nakagawa, A., Awaya, J., Tsuchya, H., Takahashi, Y. and Masuma, R. (1977). A new alkaloid AM-2282 OF Streptomyces origin. Taxonomy, fermentation, isolation and preliminary characterization. *J Antibiot (Tokyo)*, **30**, 275–282.

Oppenheim, J.J. and Yang, D. (2005). Alarmins: chemotactic activators of immune responses. *Curr Opin Immunol*, **17**, 359–365.

Paolo, N.C.D., Kalyuzhniy, O. and Shayakhmetov, D.M. (2007). Fiber shaft-chimeric adenovirus vectors lacking the KKTK motif efficiently infect liver cells in vivo. *J Virol*, **81**, 12249–12259.

Parker, A.L., Waddington, S.N., Nicol, C.G., Shayakhmetov, D.M., Buckley, S.M., Denby, L., Kemball-Cook, G., Ni, S., Lieber, A., McVey, J.H., Nicklin, S.A. and Baker, A.H. (2006). Multiple vitamin

Bibliography

K-dependent coagulation zymogens promote adenovirus-mediated gene delivery to hepatocytes. *Blood*, **108**, 2554–2561.

Parks, R., Evelegh, C. and Graham, F. (1999). Use of helper-dependent adenoviral vectors of alternative serotypes permits repeat vector administration. *Gene Ther*, **6**, 1565–1573.

Patel, M., Yanagishita, M., Roderiquez, G., Bou-Habib, D.C., Oravecz, T., Hascall, V.C. and Norcross, M.A. (1993). Cell-surface heparan sulfate proteoglycan mediates HIV-1 infection of T-cell lines. *AIDS Res Hum Retroviruses*, **9**, 167–174.

Pfaffl, M.W. (2001). A new mathematical model for relative quantification in real-time RT-PCR. *Nucleic Acids Res*, **29**, e45.

Pham, C.T., MacIvor, D.M., Hug, B.A., Heusel, J.W. and Ley, T.J. (1996). Long-range disruption of gene expression by a selectable marker cassette. *Proc Natl Acad Sci U S A*, **93**, 13090–5.

Pinkoski, M.J., Hobman, M., Heibein, J.A., Tomaselli, K., Li, F., Seth, P., Froelich, C.J. and Bleackley, R.C. (1998). Entry and trafficking of granzyme B in target cells during granzyme B-perforin-mediated apoptosis. *Blood*, **92**, 1044–54.

Pratt, W.B. and Toft, D.O. (2003). Regulation of signaling protein function and trafficking by the hsp90/hsp70-based chaperone machinery. *Exp Biol Med*, **228**, 111–33.

Provenzano, M. and Mocellin, S. (2007). Complementary techniques: validation of gene expression data by quantitative real time PCR. *Adv Exp Med Biol*, **593**, 66–73.

Puumalainen, A.M., Vapalahti, M., Agrawal, R.S., Kossila, M., Laukkanen, J., Lehtolainen, P., Viita, H., Paljärvi, L., Vanninen, R. and Ylä-Herttuala, S. (1998). Beta-galactosidase gene transfer to human malignant glioma in vivo using replication-deficient retroviruses and adenoviruses. *Hum Gene Ther*, **9**, 1769–1774.

Quan, L.T., Tewari, M., O'Rourke, K., Dixit, V., Snipas, S.J., Poirier, G.G., Ray, C., Pickup, D.J. and Salvesen, G.S. (1996). Proteolytic activation of the cell death protease Yama/CPP32 by granzyme B. *Proc Natl Acad Sci U S A*, **93**, 1972–1976.

Raja, S.M., Metkar, S.S., Höning, S., Wang, B., Russin, W.A., Pipalia, N.H., Menaa, C., Belting, M., Cao, X., Dressel, R. and Froelich, C.J. (2005). A novel mechanism for protein delivery: granzyme B undergoes electrostatic exchange from serglycin to target cells. *J Biol Chem*, **280**, 20752–61.

Raja, S.M., Wang, B., Dantuluri, M., Desai, U.R., Demeler, B., Spiegel, K., Metkar, S.S. and Froelich, C.J. (2002). Cytotoxic cell granule-mediated apoptosis. Characterization of the macromolecular complex of granzyme B with serglycin. *J Biol Chem*, **277**, 49523–49530.

Rao, L., Perez, D. and White, E. (1996). Lamin proteolysis facilitates nuclear events during apoptosis. *J Cell Biol*, **135**, 1441–1455.

Bibliography

Ravagnan, L., Gurbuxani, S., Susin, S.A., Maisse, C., Daugas, E., Zamzami, N., Mak, T., Jäättelä, M., Penninger, J.M., Garrido, C. and Kroemer, G. (2001). Heat-shock protein 70 antagonizes apoptosis-inducing factor. *Nat Cell Biol*, **3**, 839–43.

Reers, M., Smith, T.W. and Chen, L.B. (1991). J-aggregate formation of a carbocyanine as a quantitative fluorescent indicator of membrane potential. *Biochemistry*, **30**, 4480–4486.

Ricci, J.E., Muñoz-Pinedo, C., Fitzgerald, P., Bailly-Maitre, B., Perkins, G.A., Yadava, N., Scheffler, I.E., Ellisman, M.H. and Green, D.R. (2004). Disruption of mitochondrial function during apoptosis is mediated by caspase cleavage of the p75 subunit of complex I of the electron transport chain. *Cell*, **117**, 773–786.

Rogée, S., Grellier, E., Bernard, C., Colin, M. and D'Halluin, J.C. (2008). Non-heparan sulfate GAG-dependent infection of cells using an adenoviral vector with a chimeric fiber conserving its KKTK motif. *Virology*, **380**, 60–68.

Rouvier, E., Luciani, M.F. and Golstein, P. (1993). Fas involvement in Ca(2+)-independent T cell-mediated cytotoxicity. *J Exp Med*, **177**, 195–200.

Rozen, S. and Skaletsky, H. (2000). *Primer3 on the WWW for general users and for biologist programmers*. Humana Press, Totowa, NJ.

Russell, J.H. and Ley, T.J. (2002). Lymphocyte-mediated cytotoxicity. *Annu Rev Immunol*, **20**, 323–70.

Rüegg, U.T. and Burgess, G.M. (1989). Staurosporine, K-252 and UCN-01: potent but nonspecific inhibitors of protein kinases. *Trends Pharmacol Sci*, **10**, 218–220.

Saiki, R.K., Gelfand, D.H., Stoffel, S., Scharf, S.J., Higuchi, R., Horn, G.T., Mullis, K.B. and Erlich, H.A. (1988). Primer-directed enzymatic amplification of DNA with a thermostable DNA polymerase. *Science*, **239**, 487–491.

Sakahira, H., Enari, M. and Nagata, S. (1998). Cleavage of CAD inhibitor in CAD activation and DNA degradation during apoptosis. *Nature*, **391**, 96–99.

Saleh, A., Srinivasula, S.M., Balkir, L., Robbins, P.D. and Alnemri, E.S. (2000). Negative regulation of the Apaf-1 apoptosome by Hsp70. *Nat Cell Biol*, **2**, 476–83.

Samali, A., Cai, J., Zhivotovsky, B., Jones, D.P. and Orrenius, S. (1999). Presence of a pre-apoptotic complex of pro-caspase-3, Hsp60 and Hsp10 in the mitochondrial fraction of jurkat cells. *EMBO J*, **18**, 2040–8.

Samali, A. and Cotter, T.G. (1996). Heat shock proteins increase resistance to apoptosis. *Exp Cell Res*, **223**, 163–70.

Bibliography

Sandmair, A.M., Loimas, S., Puranen, P., Immonen, A., Kossila, M., Puranen, M., Hurskainen, H., Tyynelä, K., Turunen, M., Vanninen, R., Lehtolainen, P., Paljärvi, L., Johansson, R., Vapalahti, M. and Ylä-Herttuala, S. (2000). Thymidine kinase gene therapy for human malignant glioma, using replication-deficient retroviruses or adenoviruses. *Hum Gene Ther*, **11**, 2197–2205.

Santarosa, M., Favaro, D., Quaia, M. and Galligioni, E. (1997). Expression of heat shock protein 72 in renal cell carcinoma: possible role and prognostic implications in cancer patients. *Eur J Cancer*, **33**, 873–7.

Santoro, M.G. (2000). Heat shock factors and the control of the stress response. *Biochem Pharmacol*, **59**, 55–63.

Sargent, C.A., Dunham, I., Trowsdale, J. and Campbell, R.D. (1989). Human major histocompatibility complex contains genes for the major heat shock protein HSP70. *Proc Natl Acad Sci U S A*, **86**, 1968–72.

Sarin, A., Williams, M.S., Alexander-Miller, M.A., Berzofsky, J.A., Zacharchuk, C.M. and Henkart, P.A. (1997). Target cell lysis by CTL granule exocytosis is independent of ICE/Ced-3 family proteases. *Immunity*, **6**, 209–15.

Savill, J. and Fadok, V. (2000). Corpse clearance defines the meaning of cell death. *Nature*, **407**, 784–788.

Savill, J., Fadok, V., Henson, P. and Haslett, C. (1993). Phagocyte recognition of cells undergoing apoptosis. *Immunol Today*, **14**, 131–136.

Scaffidi, P., Misteli, T. and Bianchi, M.E. (2002). Release of chromatin protein HMGB1 by necrotic cells triggers inflammation. *Nature*, **418**, 191–195.

Scarlett, J.L., Sheard, P.W., Hughes, G., Ledgerwood, E.C., Ku, H.H. and Murphy, M.P. (2000). Changes in mitochondrial membrane potential during staurosporine-induced apoptosis in Jurkat cells. *FEBS Lett*, **475**, 267–272.

Schofield, K.P., Gallagher, J.T. and David, G. (1999). Expression of proteoglycan core proteins in human bone marrow stroma. *Biochem J*, **343 Pt 3**, 663–668.

Schuler, M., Herrmann, R., Greve, J.L.D., Stewart, A.K., Gatzemeier, U., Stewart, D.J., Laufman, L., Gralla, R., Kuball, J., Buhl, R., Heussel, C.P., Kommoss, F., Perruchoud, A.P., Shepherd, F.A., Fritz, M.A., Horowitz, J.A., Huber, C. and Rochlitz, C. (2001). Adenovirus-mediated wild-type p53 gene transfer in patients receiving chemotherapy for advanced non-small-cell lung cancer: results of a multicenter phase II study. *J Clin Oncol*, **19**, 1750–1758.

Seki, N., Brooks, A.D., Carter, C.R., Back, T.C., Parsoneault, E.M., Smyth, M.J., Wiltrout, R.H. and Sayers, T.J. (2002). Tumor-specific CTL kill murine renal cancer cells using both perforin and

Fas ligand-mediated lysis in vitro, but cause tumor regression in vivo in the absence of perforin. *J Immunol*, **168**, 3484–92.

Selinka, H.C., Florin, L., Patel, H.D., Freitag, K., Schmidtke, M., Makarov, V.A. and Sapp, M. (2007). Inhibition of transfer to secondary receptors by heparan sulfate-binding drug or antibody induces noninfectious uptake of human papillomavirus. *J Virol*, **81**, 10970–10980.

Sharif-Askari, E., Alam, A., Rheaume, E., Beresford, P.J., Scotto, C., Sharma, K., Lee, D., De-Wolf, W.E., Nuttall, M.E., Lieberman, J. and Sekaly, R.P. (2001). Direct cleavage of the human DNA fragmentation factor-45 by granzyme B induces caspase-activated DNase release and DNA fragmentation. *EMBO J*, **20**, 3101–13.

Shayakhmetov, D.M., Gaggar, A., Ni, S., Li, Z.Y. and Lieber, A. (2005). Adenovirus binding to blood factors results in liver cell infection and hepatotoxicity. *J Virol*, **79**, 7478–7491.

Shi, L., Keefe, D., Durand, E., Feng, H., Zhang, D. and Lieberman, J. (2005). Granzyme B binds to target cells mostly by charge and must be added at the same time as perforin to trigger apoptosis. *J Immunol*, **174**, 5456–61.

Shi, L., Mai, S., Israels, S., Browne, K., Trapani, J.A. and Greenberg, A.H. (1997). Granzyme B (GraB) autonomously crosses the cell membrane and perforin initiates apoptosis and GraB nuclear localization. *J Exp Med*, **185**, 855–66.

Shi, Y., Zheng, W. and Rock, K.L. (2000). Cell injury releases endogenous adjuvants that stimulate cytotoxic T cell responses. *Proc Natl Acad Sci U S A*, **97**, 14590–14595.

Shi, Y., Evans, J.E. and Rock, K.L. (2003). Molecular identification of a danger signal that alerts the immune system to dying cells. *Nature*, **425**, 516–521.

Short, D.M., Heron, I.D., Birse-Archbold, J.L.A., Kerr, L.E., Sharkey, J. and McCulloch, J. (2007). Apoptosis induced by staurosporine alters chaperone and endoplasmic reticulum proteins: Identification by quantitative proteomics. *Proteomics*, **7**, 3085–3096.

Shresta, S., Graubert, T.A., Thomas, D.A., Raptis, S.Z. and Ley, T.J. (1999). Granzyme A initiates an alternative pathway for granule-mediated apoptosis. *Immunity*, **10**, 595–605.

Shukla, D. and Spear, P.G. (2001). Herpesviruses and heparan sulfate: an intimate relationship in aid of viral entry. *J Clin Invest*, **108**, 503–510.

Simon, M.M., Reikerstorfer, A., Schwarz, A., Krone, C., Luger, T.A., Jaattela, M. and Schwarz, T. (1995). Heat shock protein 70 overexpression affects the response to ultraviolet light in murine fibroblasts. Evidence for increased cell viability and suppression of cytokine release. *J Clin Invest*, **95**, 926–33.

Bibliography

Smith, T., Idamakanti, N., Kylefjord, H., Rollence, M., King, L., Kaloss, M., Kaleko, M. and Stevenson, S.C. (2002). In vivo hepatic adenoviral gene delivery occurs independently of the coxsackievirus-adenovirus receptor. *Mol Ther*, **5**, 770–779.

Smith, T.A.G., Idamakanti, N., Rollence, M.L., Marshall-Neff, J., Kim, J., Mulgrew, K., Nemerow, G.R., Kaleko, M. and Stevenson, S.C. (2003). Adenovirus serotype 5 fiber shaft influences in vivo gene transfer in mice. *Hum Gene Ther*, **14**, 777–787.

Smyth, M.J., Thia, K.Y., Street, S.E., MacGregor, D., Godfrey, D.I. and Trapani, J.A. (2000). Perforin-mediated cytotoxicity is critical for surveillance of spontaneous lymphoma. *J Exp Med*, **192**, 755–60.

Sreedhar, A.S. and Csermely, P. (2004). Heat shock proteins in the regulation of apoptosis: new strategies in tumor therapy: a comprehensive review. *Pharmacol Ther*, **101**, 227–57.

Srivastava, P. (2002a). Interaction of heat shock proteins with peptides and antigen presenting cells: chaperoning of the innate and adaptive immune responses. *Annu Rev Immunol*, **20**, 395–425.

Srivastava, P. (2002b). Roles of heat-shock proteins in innate and adaptive immunity. *Nat Rev Immunol*, **2**, 185–94.

Srivastava, P.K., Menoret, A., Basu, S., Binder, R.J. and McQuade, K.L. (1998). Heat shock proteins come of age: primitive functions acquire new roles in an adaptive world. *Immunity*, **8**, 657–65.

Srivastava, P.K., Udono, H., Blachere, N.E. and Li, Z. (1994). Heat shock proteins transfer peptides during antigen processing and CTL priming. *Immunogenetics*, **39**, 93–8.

Stahnke, K., Mohr, A., Liu, J., Meyer, L.H., Karawajew, L. and Debatin, K.M. (2004). Identification of deficient mitochondrial signaling in apoptosis resistant leukemia cells by flow cytometric analysis of intracellular cytochrome c, caspase-3 and apoptosis. *Apoptosis*, **9**, 457–465.

Stankiewicz, A.R., Lachapelle, G., Foo, C.P., Radicioni, S.M. and Mosser, D.D. (2005). Hsp70 inhibits heat-induced apoptosis upstream of mitochondria by preventing Bax translocation. *J Biol Chem*, **280**, 38729–39.

Stark, S. (1995). *Molekulargenetische Charakterisierung von Genen, deren Expression nach Hitzeschock in einer Myelomlinie von Rattus norvegicus verändert ist*. Cuvillier Verlag.

Stevens, R.L., Avraham, S., Gartner, M.C., Bruns, G.A., Austen, K.F. and Weis, J.H. (1988). Isolation and characterization of a cDNA that encodes the peptide core of the secretory granule proteoglycan of human promyelocytic leukemia HL-60 cells. *J Biol Chem*, **263**, 7287–7291.

Stinchcombe, J.C., Bossi, G., Booth, S. and Griffiths, G.M. (2001). The immunological synapse of CTL contains a secretory domain and membrane bridges. *Immunity*, **15**, 751–61.

Bibliography

Stinchcombe, J.C. and Griffiths, G.M. (2003). The role of the secretory immunological synapse in killing by CD8+ CTL. *Semin Immunol*, **15**, 301–5.

Strasser, A. and Pellegrini, M. (2004). T-lymphocyte death during shutdown of an immune response. *Trends Immunol*, **25**, 610–615.

Sugawara, S., Nowicki, M., Xie, S., Song, H.J. and Dennert, G. (1990). Effects of stress on lysability of tumor targets by cytotoxic T cells and tumor necrosis factor. *J Immunol*, **145**, 1991–8.

Sukal, S. and Leyh, T.S. (2001). Product release during the first turnover of the ATP sulfurylase-GTPase. *Biochemistry*, **40**, 15009–16.

Sutton, V.R., Davis, J.E., Cancilla, M., Johnstone, R.W., Ruefli, A.A., Sedelies, K., Browne, K.A. and Trapani, J.A. (2000). Initiation of apoptosis by granzyme B requires direct cleavage of bid, but not direct granzyme B-mediated caspase activation. *J Exp Med*, **192**, 1403–14.

Tang, D. and Kidd, V.J. (1998). Cleavage of DFF-45/ICAD by multiple caspases is essential for its function during apoptosis. *J Biol Chem*, **273**, 28549–28552.

Tang, D., Lahti, J.M. and Kidd, V.J. (2000). Caspase-8 activation and bid cleavage contribute to MCF7 cellular execution in a caspase-3-dependent manner during staurosporine-mediated apoptosis. *J Biol Chem*, **275**, 9303–9307.

Taylor, R.C., Cullen, S.P. and Martin, S.J. (2008). Apoptosis: controlled demolition at the cellular level. *Nat Rev Mol Cell Biol*, **9**, 231–241.

Thomas, D.A., Du, C., Xu, M., Wang, X. and Ley, T.J. (2000). DFF45/ICAD can be directly processed by granzyme B during the induction of apoptosis. *Immunity*, **12**, 621–32.

Thériault, J.R., Adachi, H. and Calderwood, S.K. (2006). Role of scavenger receptors in the binding and internalization of heat shock protein 70. *J Immunol*, **177**, 8604–8611.

Todryk, S., Melcher, A.A., Hardwick, N., Linardakis, E., Bateman, A., Colombo, M.P., Stoppacciaro, A. and Vile, R.G. (1999). Heat shock protein 70 induced during tumor cell killing induces Th1 cytokines and targets immature dendritic cell precursors to enhance antigen uptake. *J Immunol*, **163**, 1398–408.

Tomko, R.P., Xu, R. and Philipson, L. (1997). HCAR and MCAR: the human and mouse cellular receptors for subgroup C adenoviruses and group B coxsackieviruses. *Proc Natl Acad Sci U S A*, **94**, 3352–3356.

Towbin, H., Staehelin, T. and Gordon, J. (1979). Electrophoretic transfer of proteins from polyacrylamide gels to nitrocellulose sheets: procedure and some applications. *Proc Natl Acad Sci U S A*, **76**, 4350–4354.

Bibliography

Tran, S.E., Meinander, A., Holmstrom, T.H., Rivero-Muller, A., Heiskanen, K.M., Linnau, E.K., Courtney, M.J., Mosser, D.D., Sistonen, L. and Eriksson, J.E. (2003). Heat stress downregulates FLIP and sensitizes cells to Fas receptor-mediated apoptosis. *Cell Death Differ*, **10**, 1137–47.

Trapani, J.A., Jans, D.A., Jans, P.J., Smyth, M.J., Browne, K.A. and Sutton, V.R. (1998). Efficient nuclear targeting of granzyme B and the nuclear consequences of apoptosis induced by granzyme B and perforin are caspase-dependent, but cell death is caspase-independent. *J Biol Chem*, **273**, 27934–8.

Trapani, J.A. and Smyth, M.J. (2002). Functional significance of the perforin/granzyme cell death pathway. *Nat Rev Immunol*, **2**, 735–47.

Trapani, J.A., Sutton, V.R., Thia, K.Y., Li, Y.Q., Froelich, C.J., Jans, D.A., Sandrin, M.S. and Browne, K.A. (2003). A clathrin/dynamin- and mannose-6-phosphate receptor-independent pathway for granzyme B-induced cell death. *J Cell Biol*, **160**, 223–233.

Trautinger, F., Kokesch, C., Klosner, G., Knobler, R.M. and Kindas-Mugge, I. (1999). Expression of the 72-kD heat shock protein is induced by ultraviolet A radiation in a human fibrosarcoma cell line. *Exp Dermatol*, **8**, 187–92.

Trieb, K., Lechleitner, T., Lang, S., Windhager, R., Kotz, R. and Dirnhofer, S. (1998). Heat shock protein 72 expression in osteosarcomas correlates with good response to neoadjuvant chemotherapy. *Hum Pathol*, **29**, 1050–5.

Twomey, C. and McCarthy, J.V. (2005). Pathways of apoptosis and importance in development. *J Cell Mol Med*, **9**, 345–359.

Udono, H. and Srivastava, P.K. (1993). Heat shock protein 70-associated peptides elicit specific cancer immunity. *J Exp Med*, **178**, 1391–6.

Van Parijs, L. and Abbas, A.K. (1996). Role of Fas-mediated cell death in the regulation of immune responses. *Curr Opin Immunol*, **8**, 355–61.

Vega, V.L., Rodríguez-Silva, M., Frey, T., Gehrmann, M., Diaz, J.C., Steinem, C., Multhoff, G., Arispe, N. and Maio, A.D. (2008). Hsp70 translocates into the plasma membrane after stress and is released into the extracellular environment in a membrane-associated form that activates macrophages. *J Immunol*, **180**, 4299–4307.

Veugelers, K., Motyka, B., Frantz, C., Shostak, I., Sawchuk, T. and Bleackley, R.C. (2004). The granzyme B-serglycin complex from cytotoxic granules requires dynamin for endocytosis. *Blood*, **103**, 3845–53.

Veugelers, K., Motyka, B., Goping, I.S., Shostak, I., Sawchuk, T. and Bleackley, R.C. (2006). Granule-mediated killing by granzyme B and perforin requires a mannose 6-phosphate receptor and is augmented by cell surface heparan sulfate. *Mol Biol Cell*, **17**, 623–33.

Bibliography

Waddington, S.N., McVey, J.H., Bhella, D., Parker, A.L., Barker, K., Atoda, H., Pink, R., Buckley, S.M.K., Greig, J.A., Denby, L., Custers, J., Morita, T., Francischetti, I.M.B., Monteiro, R.Q., Barouch, D.H., van Rooijen, N., Napoli, C., Havenga, M.J.E., Nicklin, S.A. and Baker, A.H. (2008). Adenovirus serotype 5 hexon mediates liver gene transfer. *Cell*, **132**, 397–409.

Walters, R.W., Freimuth, P., Moninger, T.O., Ganske, I., Zabner, J. and Welsh, M.J. (2002). Adenovirus fiber disrupts CAR-mediated intercellular adhesion allowing virus escape. *Cell*, **110**, 789–799.

Waterhouse, N.J., Sedelies, K.A., Sutton, V.R., Pinkoski, M.J., Thia, K.Y., Johnstone, R., Bird, P.I., Green, D.R. and Trapani, J.A. (2006a). Functional dissociation of DeltaPsim and cytochrome c release defines the contribution of mitochondria upstream of caspase activation during granzyme B-induced apoptosis. *Cell Death Differ*, **13**, 607–618.

Waterhouse, N.J., Sutton, V.R., Sedelies, K.A., Ciccone, A., Jenkins, M., Turner, S.J., Bird, P.I. and Trapani, J.A. (2006b). Cytotoxic T lymphocyte-induced killing in the absence of granzymes A and B is unique and distinct from both apoptosis and perforin-dependent lysis. *J Cell Biol*, **173**, 133–44.

Weksberg, R., Hughes, S., Moldovan, L., Bassett, A.S., Chow, E.W.C. and Squire, J.A. (2005). A method for accurate detection of genomic microdeletions using real-time quantitative PCR. *BMC Genomics*, **6**, 180.

Wells, A.D., Rai, S.K., Salvato, M.S., Band, H. and Malkovsky, M. (1998). Hsp72-mediated augmentation of MHC class I surface expression and endogenous antigen presentation. *Int Immunol*, **10**, 609–17.

Wickham, T.J., Mathias, P., Cheresh, D.A. and Nemerow, G.R. (1993). Integrins alpha v beta 3 and alpha v beta 5 promote adenovirus internalization but not virus attachment. *Cell*, **73**, 309–319.

Willis, S.N. and Adams, J.M. (2005). Life in the balance: how BH3-only proteins induce apoptosis. *Curr Opin Cell Biol*, **17**, 617–625.

Willis, S.N., Fletcher, J.I., Kaufmann, T., van Delft, M.F., Chen, L., Czabotar, P.E., Ierino, H., Lee, E.F., Fairlie, W.D., Bouillet, P., Strasser, A., Kluck, R.M., Adams, J.M. and Huang, D.C.S. (2007). Apoptosis initiated when BH3 ligands engage multiple Bcl-2 homologs, not Bax or Bak. *Science*, **315**, 856–859.

Worgall, S., Wolff, G., Falck-Pedersen, E. and Crystal, R.G. (1997). Innate immune mechanisms dominate elimination of adenoviral vectors following in vivo administration. *Hum Gene Ther*, **8**, 37–44.

Wurst, W., Benesch, C., Drabent, B., Rothermel, E., Benecke, B.J. and Günther, E. (1989). Localization of heat shock protein 70 genes inside the rat major histocompatibility complex close to class III genes. *Immunogenetics*, **30**, 46–9.

Bibliography

Wyllie, A.H., Kerr, J.F. and Currie, A.R. (1980). Cell death: the significance of apoptosis. *Int Rev Cytol*, **68**, 251–306.

Xanthoudakis, S., Roy, S., Rasper, D., Hennessey, T., Aubin, Y., Cassady, R., Tawa, P., Ruel, R., Rosen, A. and Nicholson, D.W. (1999). Hsp60 accelerates the maturation of pro-caspase-3 by upstream activator proteases during apoptosis. *EMBO J*, **18**, 2049–56.

Yamagishi, N., Ishihara, K., Saito, Y. and Hatayama, T. (2006). Hsp105 family proteins suppress staurosporine-induced apoptosis by inhibiting the translocation of Bax to mitochondria in HeLa cells. *Exp Cell Res*, **312**, 3215–3223.

Yang, X., Stennicke, H.R., Wang, B., Green, D.R., Jänicke, R.U., Srinivasan, A., Seth, P., Salvesen, G.S. and Froelich, C.J. (1998). Granzyme B mimics apical caspases. Description of a unified pathway for trans-activation of executioner caspase-3 and -7. *J Biol Chem*, **273**, 34278–34283.

Yoshida, H., Kong, Y.Y., Yoshida, R., Elia, A.J., Hakem, A., Hakem, R., Penninger, J.M. and Mak, T.W. (1998). Apaf1 is required for mitochondrial pathways of apoptosis and brain development. *Cell*, **94**, 739–750.

Youle, R.J. (2007). Cell biology. Cellular demolition and the rules of engagement. *Science*, **315**, 776–777.

Youle, R.J. and Strasser, A. (2008). The BCL-2 protein family: opposing activities that mediate cell death. *Nat Rev Mol Cell Biol*, **9**, 47–59.

Yue, T.L., Wang, C., Romanic, A.M., Kikly, K., Keller, P., DeWolf, W.E., Hart, T.K., Thomas, H.C., Storer, B., Gu, J.L., Wang, X. and Feuerstein, G.Z. (1998). Staurosporine-induced apoptosis in cardiomyocytes: A potential role of caspase-3. *J Mol Cell Cardiol*, **30**, 495–507.

Yun, C.O., Yoon, A.R., Yoo, J.Y., Kim, H., Kim, M., Ha, T., Kim, G.E., Kim, H. and Kim, J.H. (2005). Coxsackie and adenovirus receptor binding ablation reduces adenovirus liver tropism and toxicity. *Hum Gene Ther*, **16**, 248–261.

Zajac, A.J., Dye, J.M. and Quinn, D.G. (2003). Control of lymphocytic choriomeningitis virus infection in granzyme B deficient mice. *Virology*, **305**, 1–9.

Zhang, D., Pasternack, M.S., Beresford, P.J., Wagner, L., Greenberg, A.H. and Lieberman, J. (2001). Induction of rapid histone degradation by the cytotoxic T lymphocyte protease Granzyme A. *J Biol Chem*, **276**, 3683–90.

Zhang, X.D., Gillespie, S.K. and Hersey, P. (2004). Staurosporine induces apoptosis of melanoma by both caspase-dependent and -independent apoptotic pathways. *Mol Cancer Ther*, **3**, 187–197.

Zimmermann, K.C. and Green, D.R. (2001). How cells die: apoptosis pathways. *J Allergy Clin Immunol*, **108**, S99–103.

List of Figures

2.1	Apoptosis	10
2.2	Role of HSP70 in the stress response and in the immune system	18
4.1	RNA quality analysis with a pico chip	55
4.2	Setup of the microarray	56
4.3	Generation of labelled cRNA for a two-colour microarray	57
4.4	Calculation of the primer efficiency from the slope of cDNA dilutions for the *Hsp70* primer	64
5.1	Permanent HSP70 overexpression in the Ge-Hsp70-D clone	68
5.2	Acute HSP70 overexpression in Ge-tet-1 and 2 clones	69
5.3	HSP70 expression in cells for RNA isolation	70
5.4	MA-plots of Ge-tra and Ge-tet-1 cells	71
5.5	Changes in expression of 13 selected genes as determined by a whole human microarray analysis and by quantitative real-time PCR	73
5.6	Flow cytometric analysis of annexin V/PI staining	75
5.7	Acute HSP70 overexpression did not influence GrB-induced apoptosis as measured by binding of annexin to exposed phosphatidylserines	76
5.8	Flow cytometric analysis of the sub G1-peak	77
5.9	Acute HSP70 overexpression can enhance GrB-induced apoptosis as determined by sub G1-peak analysis	78
5.10	Permanent overexpression of HSP70 does not influence GrB-induced apoptosis as determined by sub G1-peak analysis	79
5.11	Acute HSP70 overexpression does not influence GrB-uptake	81
5.12	Acute HSP70 overexpression partially protects from staurosporine-induced apoptosis	82
5.13	Acute HSP70 overexpression does not improve staurosporine-induced apoptosis as determined by sub G1-peak analysis	84
5.14	Permanent overexpression of HSP70 does not influence staurosporine-induced apoptosis as determined by sub G1-peak analysis	85

List of Figures

5.15 Analysed key steps in apoptosis 86
5.16 Determining the mitochondrial membrane potential $\Delta\Psi$ by flow cytometry ... 87
5.17 Acute HSP70 overexpression did not influence the change in mitochondrial membrane-potential $\Delta\Psi$ after GrB-induced apoptosis 88
5.18 Acute HSP70 overexpression did not influence the change in mitochondrial membrane potential $\Delta\Psi$ after staurosporine-induced apoptosis 89
5.19 Flow cytometric analysis of the release of cytochrome c from mitochondria ... 90
5.20 Acute HSP70 overexpression does not significantly increase the release of cytochrome c from mitochondria after GrB-induced apoptosis 91
5.21 Acute HSP70 overexpression does not protect Ge-tet-1 cells from the release of cytochrome c from mitochondria after staurosporine-induced apoptosis 92
5.22 Staurosporine-induced apoptosis did not change caspase-8 levels 93
5.23 Acute HSP70 overexpression did not influence the activation of caspase-3 by GrB-induced apoptosis 94
5.24 Acute HSP70 overexpression significantly decreased the percentage of cells with active caspase-3 by staurosporine-induced apoptosis 95
5.25 Flow cytometric analysis of the activation of caspase-3 in NK cell-mediated apoptosis 96
5.26 Acute HSP70 overexpression did not change the activation of caspase-3 by NK cell-induced apoptosis 97
5.27 Acute HSP70 overexpression did not induce DNA fragmentation as detected by laddering of DNA 98
5.28 Uptake of Alexa488-labelled GrB into *Sulf*MEFs 99
5.29 Lysis of *Sulf* Dko by OT-I-derived CTLs is slightly higher than the lysis of Wt MEFs 101
5.30 $Sulf2^{-/-}$ cells are as susceptible to antigen-specific CTLs as Wt cells 102
5.31 Lysis of $Sulf1^{-/-}$ cells by OT-I-derived CTLs is higher than of Wt cells 103
5.32 CTL-induced apoptosis does not seem to be changed in *Sulf* Dko cells compared to Wt cells 105
5.33 CTL-induced apoptosis is not different in $Sulf2^{-/-}$ and Wt cells 106
5.34 Apoptosis induced by CTLs is not different in $Sulf1^{-/-}$ compared to Wt MEFs . 107
5.35 Determining H2K^b expression on Wt and *Sulf*-deficient cells 108
5.36 Determination of H2K^b expression on Wt and *Sulf*-deficient cells in [^{3}H]-Thymidine-release assays 109
5.37 Adenoviral delivery of GrB into MEFs shows slightly reduced levels of apoptosis in the $Sulf2^{-/-}$ and the *Sulf* double-deficient cells compared to Wt cells 110

List of Figures

5.38 GFP expression as result of the productive infection of AdV-GFP is increased in *Sulf*-deficient clones in comparison to Wt cells 111
5.39 CAR expression levels are elevated in *Sulf* Dko in comparison to Wt and $Sulf1^{-/-}$ and $Sulf2^{-/-}$ clones 112
5.40 Integrin $α_v$ content on the different *Sulf* clones 113
5.41 Heparinase II and III-treated *Sulf* MEFs are positive for HS stubs detected with the 3G10 antibody 114
5.42 Heparinase II and III treatment of *Sulf* MEFs increases productive infection with AdV 115
5.43 CHO mutant A745 does not express HSPGs 116
5.44 Uptake of AdV into CHO and A745 116

6.1 Scheme of adenovirus 132
6.2 Model of AdV uptake into cells 134

A.1 Dissociation curve for the *Hsp70* primer 173

List of Tables

3.1 Antibodies... 21
3.2 Secondary antibodies and isotype controls... 22
3.3 Dyes... 22
3.4 Primers... 23
3.5 Chemicals... 24
3.6 Kits... 27
3.7 Buffers and stock solutions... 27
3.8 Buffers and solutions for microarray... 30
3.9 Cell lines... 31
3.10 Media... 33
3.11 Laboratory animals... 33
3.12 Used laboratory equipment... 34
3.13 Disposable plastic ware and other disposables... 36
3.14 Providers... 37

4.1 Master mix dsDNA for microarray... 56
4.2 Master mix for generating labelled cRNA for microarray... 58
4.3 Fragmentation mix for one 4× 44K microarray... 59
4.4 Master mix reverse transcription... 61
4.5 Standard PCR reaction... 61
4.6 Standard PCR programme... 61
4.7 Master mix quantitative real-time PCR... 63
4.8 Quantitative real-time PCR programme... 63

5.1 Analysis of microarray data... 70
5.2 Selected genes for validation of microarray data by quantitative real-time PCR... 72

B.1 Quantification of labelled amplified cRNA for microarray... 175
B.2 All genes found to be regulated in micoarray analysis... 176

Acknowledgements

I would like to thank Prof. Dr. Detlef Doenecke and Prof. Dieter Heineke for being official reviewers of my thesis. I am furthermore thankful to PD Dr. Lutz Walter, PD Dr. Michael Hoppert, Prof. Dr. Sigrid Hoyer-Fender, and Prof. Dr. Peter Kappeler, who completed the committee for the evaluation of my thesis.

I would like to thank my supervisor PD. Dr. Ralf Dressel for giving me the opportunity to work on the topic of HSP70 and sulphatases 1 and 2 and their role in apoptosis. Although, he should actually never had employed me, as he explained to me a few weeks ago with a friendly grin on his face referring to (Djerassi 1993), I am very thankful to complete my doctoral thesis in his lab. He is always open for discussions, has great ideas, and is always very accurate and reliable.

Leslie Elsner is the best technical assistant I have ever known. Not only for her practical skills and insider tricks using our laboratory equipment, but also for her personality, for showing me that there is a life outside science, and for the impressive demonstration that once one does enough sports, one can eat as much chocolate as one wants ;) Thanks.

My 3 years of laboratory work, reading papers, writing abstracts, proof-reading abstracts, designing posters and primers, and finally writing the thesis, and many other things, would not have been as efficient, well structured, funny, informative, and excited if there would not have been my nice colleagues Dr. Vijayakumar "VJ" Muppala, Dr. Peter Novota, and my special Kinderzimmerkollegin Miriam "Wie konnte das denn passieren" Ensslen. Thank you for always being there for me.

Furthermore, I want to thank the GRK1034 for letting me participate in their educational programme. I learned a lot. I am very thankful to all other doctoral students from the GRK1034 but especially Dr. Ines Ecke and Lars Böckmann for sharing and discussing scientific problems but also feelings with me, one can just understand while doing a scientific doctoral thesis.

To avoid that water bombs are thrown at me when I will kiss the Gänseliesl, I also need to thank Dr. Gabriela Salinas-Riester and Lennart Opitz for their extraordinary help while performing the microarray analysis and the quantitative real-time PCR. I kept my promise!

I additionally want to thank Prof. Dr. Thomas Dierks for providing us with the *Sulf* MEFs, Prof. Dr. Christopher Froelich for giving us the labelled GrB, Dr. Tim Seidler for allowing us to use his AdV and to work in the S2-laboratory, Anne Schmöle for organising papers I could

Acknowledgements

not get, and Dr. Fridtjof Nußbeck for the statistical analysis using ANOVA.

Apart from that, I am deeply grateful to my best friends Franziska Jurk and Stephan Janz. Without your mental and technical support, lending me your ear for all my problems and giving me a shoulder to cry on, encouraging me to go on but also distracting me if neccessary, and making me happy, this thesis might not have been finished, at least not with having in mind that I am not alone. In short, thanks for your friendship.

I am of course also very grateful to my parents Bedri and Marion Demiroglu, who tried to support me always as good as they could, which is particularly not as easy due to the distance between Cologne and Göttingen, but I know for sure, that your thoughts were always with me.

Last but not least I want to thank my boyfriend Gunnar Nußbeck. I know that doing a doctoral thesis was not an easy thing for me, but it must have been even harder for you. All my emotions, excitement, frustration, anger, stress, and happiness strongly correlated with the results of my experiments, but nevertheless you were always there for me. You gave me the energy and strength to complete this thesis, you teached me to be a "tank-rhino", which was not always that easy, and you always believed in me. Thanks, I love you.

This work was financed by the DFG grant DR 394/2-3 and the DFG graduate college 1034: Genetic polymorphisms in Oncology.

Appendix A

Quantitative real-time PCR

A.1 Dissociation curves

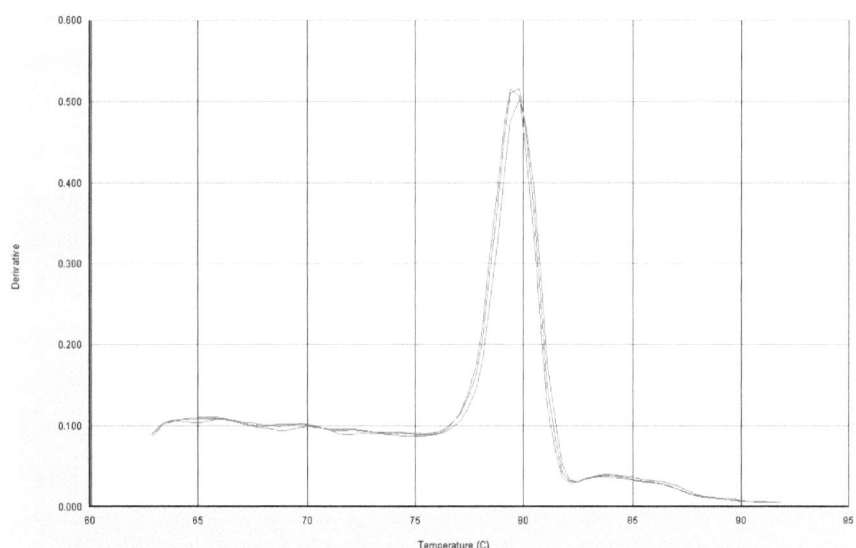

Figure A.1: Dissociation curve for the *Hsp70* primer Depicted is a screenshot of the dissociation curves of the cDNA of Ge-tet-2 triplicate amplified with the *Hsp70* primer. The x-axis shows the temperature in °C and the y-axis the derivative of the raw data of the SYBR green fluorescence. The dissociation curve is a quality control for the qRT-PCR, as it ensures an equal amplification of the products with one primer, if all curves overlap, as it is the case here.

Appendix B

Microarray

B.1 Quantification of labelled amplified cRNA for microarray

Table B.1: Quantification of labelled amplified cRNA for microarray

sample	name	cRNA conc. [ng/μl]	cRNA yield [μg]	concentration of Cy3 or Cy5 [pmole/μl]	specific activity [pmole/μg]
1-Cy3	Ge-tra	695.4	20.86	7.6	10.93
1-Cy5	Ge-tra	404.2	12.13	6.0	14.84
2-Cy3	Ge-tra +dox	587.0	17.61	6.2	10.56
2-Cy5	Ge-tra +dox	708.3	21.25	16.3	23.01
3-Cy3	Ge-tet-1	427.7	12.83	4.5	10.52
3-Cy5	Ge-tet-1	633.5	19.01	11.3	17.84
4-Cy3	Ge-tet-1 +dox	426.5	12.80	4.0	9.38
4-Cy5	Ge-tet-1 +dox	479.7	14.39	8.8	18.34

Appendix B Microarray

B.2 All genes found to be regulated in Ge-tra and Ge-tet-1 upon treatment with doxycycline by microarray analysis

Table B.2: All genes found to be regulated in micoarray analysis Regulated genes with log 2-fold change in expression of smaller than -2 or larger than 2, are listed with gene abbreviation (Abbrev.), gene name (Name), log 2-fold change in expression in Ge-tra versus Ge-tra plus doxycycline (Expr. tra) and Ge-tet versus Ge-tet plus doxycycline (Expr. tet), and the respective p-values (p-val.). Negative values in the change in expression mean an upregulation of the gene upon doxycycline treatment and positive values a downregulation. If for one transcript more than one oligo was present on the microarray all values of expression and p-values were given seperately for the single oligos. The fold change in expression can be calculated from the log 2-fold change in expression by using formula 4.10 on page 65. p-values are significant with a type I error rate α of 0.01. Significant expression and p-values are printed in bold letters.

Abbrev.	Name	Expr. tra	p-val.	Expr. tet	p-val.
ANGPTL4	Homo sapiens angiopoietin-like 4, transcript variant 1	2.86	**1.21E-04**	0.53	1.00E+00
SPON2	Homo sapiens spondin 2	2.42	**1.20E-06**	1.19	**1.11E-05**
A_24_P852099	Unknown	2.26	**1.56E-05**	1.17	**1.32E-04**
CCL2	Homo sapiens chemokine (C-C motif) ligand 2	2.22	**1.74E-04**	0.21	1.00E+00
BFSP1	Homo sapiens beaded filament structural protein 1, filensin	2.22	**6.77E-05**	1.30	**5.60E-04**
COX1	Cytochrome c oxidase subunit 1 (EC 1.9.3.1) (Cytochrome c oxidase polypeptide I).	2.20	**3.16E-05**	0.98	1.00E+00
NFE2	Homo sapiens nuclear factor (erythroid-derived 2)	2.17	**1.17E-04**	0.62	1.00E+00
BCL10	B-cell lymphoma 10	-0.58	1.00E+00	-2.12	**5.55E-04**
CASP8	caspase 8, apoptosis-related cysteine peptidase	-2.02	**2.05E-04**	4.48	**3.06E-05**
XBP1	Homo sapiens X-box binding protein 1, transcript variant 1	-2.02	**7.55E-05**	-0.14	1.00E+00
CARS	Homo sapiens cysteinyl-tRNA synthetase, transcript variant 4	-2.04	**5.72E-05**	-0.65	1.00E+00
PHGDH	Homo sapiens phosphoglycerate dehydrogenase	-2.09	**2.70E-05**	-1.41	**1.17E-04**
ND6	NADH-ubiquinone oxidoreductase chain 6 (EC 1.6.5.3) (NADH dehydrogenase subunit 6).	-2.11	**3.12E-03**	-1.13	3.16E-02
HOXB9	Homo sapiens homeobox B9	-2.18	**5.55E-05**	-1.31	**4.12E-04**
LOC643159	PREDICTED: Homo sapiens similar to activating transcription factor 4, transcript variant 1	-2.18	**5.35E-05**	-1.42	2.88E-04
A_24_P222054	Homo sapiens DC48 mRNA, complete cds	-2.22	**3.74E-05**	-1.10	4.78E-04
ATF4	Homo sapiens activating transcription factor 4 (tax-responsive enhancer element B67), transcript variant 1	-2.22	**1.60E-06**	-1.32	**1.10E-05**
TRIB3	Homo sapiens tribbles homolog 3 (Drosophila)	-2.28	**1.16E-03**	-0.35	1.00E+00
ALDH1L2	Homo sapiens mRNA; cDNA DKFZp686A16126	-2.32	**1.45E-05**	-1.19	**1.17E-04**
LOC732249	PREDICTED: Homo sapiens similar to phosphoserine aminotransferase isoform 1	-2.36	**1.10E-04**	-0.72	1.00E+00
GARS	Homo sapiens glycyl-tRNA synthetase	-2.37	**4.32E-05**	-1.26	**4.69E-04**
CBS	Homo sapiens cystathionine-beta-synthase	-2.39	**2.38E-04**	-1.35	2.41E-03
S100P	Homo sapiens S100 calcium binding protein P	-2.43	**5.25E-05**	-1.32	**5.31E-04**
DDIT3	Homo sapiens DNA-damage-inducible transcript 3	-2.43	**2.70E-05**	-0.56	1.00E+00
DDIT4	Homo sapiens DNA-damage-inducible transcript 4	-2.46	**2.64E-05**	-1.27	2.52E-04
CBS	Homo sapiens cystathionine-beta-synthase	-2.47	**3.74E-05**	-1.18	**5.48E-04**

Continued on next page

B.2 All genes found to be regulated in micoarray analysis

Table B.2 – continued from previous page

Abbrev.	Name	Expr. tra	p-val.	Expr. tet	p-val.
CEBPG	Homo sapiens CCAAT/enhancer binding protein (C/EBP), gamma	-2.48	7.87E-05	-1.43	6.94E-04
JDP2	Homo sapiens jun dimerization protein 2	-2.51	4.81E-04	-0.94	1.00E+00
STC2	Homo sapiens stanniocalcin 2	-2.51	1.06E-04	-1.12	2.39E-03
A_24_P25040	Unknown	-2.54	1.91E-03	-1.83	7.31E-03
A_32_P50670	AGENCOURT_6552906 NIH_MGC_85 Homo sapiens cDNA clone IMAGE:5552012 5'	-2.66	1.07E-05	-1.38	8.00E-05
ATF4	Homo sapiens activating transcription factor 4, transcript variant 1	-2.68	5.14E-03	-0.46	1.00E+00
MTHFD2	Homo sapiens methylenetetrahydrofolate dehydrogenase (NADP+ dependent) 2, methenyltetrahydrofolate cyclohydrolase, nuclear gene encoding mitochondrial protein, transcript variant 1	-2.70	5.00E-06	-1.29	3.39E-05
STC2	Homo sapiens stanniocalcin 2	-2.76	3.55E-05	-0.90	1.00E+00
VLDLR	Homo sapiens very low density lipoprotein receptor, transcript variant 1	-2.76	1.45E-05	-1.46	1.01E-04
LOC651255	PREDICTED: Homo sapiens similar to phosphoserine aminotransferase isoform 2	-3.00	3.65E-05	-0.98	1.00E+00
TRIB3	Homo sapiens tribbles homolog 3 (Drosophila)	-3.07	7.05E-05	-0.80	1.00E+00
A_32_P158376	Unknown	-3.08	7.25E-05	-1.59	9.06E-04
GRB10	Homo sapiens growth factor receptor-bound protein 10, transcript variant 4	-3.09	1.46E-05	-1.44	1.57E-04
A_24_P802562	Unknown	-3.26	1.41E-05	-1.93	7.53E-05
A_24_P699737	Homo sapiens cDNA FLJ12277 fis, clone MAMMA1001711.	-3.31	6.70E-06	-1.33	9.69E-05
PSAT1	Homo sapiens phosphoserine aminotransferase 1, transcript variant 1	-3.41	2.64E-05	-1.31	6.79E-04
A_24_P924185	RC1=NADH dehydrogenase subunit 3 homolog/ND3 homolog	-3.45	9.00E-06	-1.79	6.69E-05
GRB10	Homo sapiens growth factor receptor-bound protein 10, transcript variant 4	-3.66	2.64E-05	-0.84	1.00E+00
NUPR1	Homo sapiens nuclear protein 1, transcript variant 2	-4.02	2.98E-04	-1.76	7.30E-03
CTH	Homo sapiens cystathionase (cystathionine gamma-lyase), transcript variant 1	-4.06	1.35E-04	-1.90	2.54E-03
ALDH1L2	Homo sapiens mRNA; cDNA DKFZp686A16126	-4.31	1.02E-05	-1.96	1.02E-04
ASNS	asparagine synthetase	-4.45	3.74E-05	-2.12	5.57E-04
ASNS	asparagine synthetase	-4.51	5.54E-05	-2.14	8.78E-04
A_24_P614579	Homo sapiens, clone IMAGE:3350658	-4.51	1.00E-07	-1.88	2.40E-06
HSPA1A	heat shock protein 1A 70 kDa	1.03	8.01E-05	-8.81	4.00E-07
HSPA1A	heat shock protein 1A 70 kDa	0.96	1.00E+00	-7.49	1.10E-06
TXNRD1	thioredoxin reductase 1	-0.66	1.00E+00	-4.87	4.50E-04
ARPP-19	cyclic AMP phosphoprotein, 19 kD	0.02	1.00E+00	-3.92	1.07E-03
SAPS3	SAPS domain family, member 3	0.79	1.00E+00	-3.79	6.69E-04
EIF5	eukaryotic translation initiation factor 5	-0.57	1.00E+00	-3.71	1.47E-05
AFF1	AF4/FMR2 family, member 1	-0.43	1.00E+00	-3.31	1.13E-03
TMED7	transmembrane emp24 protein transport domain containing 7	-0.58	1.00E+00	-3.28	6.34E-03
FAM84B	family with sequence similarity 84, member B	-0.28	1.00E+00	-3.23	1.18E-03
RPS18	ribosomal protein S18	0.20	1.00E+00	-3.12	6.47E-05
A_32_P202708	-	-0.72	1.00E+00	-3.12	5.46E-04

Continued on next page

Appendix B Microarray

Table B.2 – continued from previous page

Abbrev.	Name	Expr. tra	p-val.	Expr. tet	p-val.
SSR1	signal sequence receptor, alpha (translocon-associated protein alpha)	0.14	1.00E+00	**-3.09**	**1.16E-03**
NPTN	neuroplastin	-0.71	1.00E+00	**-3.06**	**2.40E-04**
TBL1Y	transducin (beta)-like 1Y-linked	-1.26	1.41E-03	**-2.88**	**1.15E-04**
RCC2	regulator of chromosome condensation 2	-0.25	1.00E+00	**-2.83**	**8.04E-04**
BAZ1B	bromodomain adjacent to zinc finger domain, 1B	0.08	1.00E+00	**-2.80**	**2.53E-05**
HP1BP3	heterochromatin protein 1, binding protein 3	0.37	1.00E+00	**-2.75**	**4.91E-04**
MTAP	methylthioadenosine phosphorylase	-0.52	1.00E+00	**-2.70**	**8.53E-04**
CSNK1G3	casein kinase 1, gamma 3	-0.65	1.00E+00	**-2.69**	**2.38E-03**
CIT	citron (rho-interacting, serine/threonine kinase 21)	-0.62	1.00E+00	**-2.66**	**7.78E-05**
HIST1H4C	histone cluster 1, H4c	-1.00	1.38E-01	**-2.64**	**7.34E-03**
VASP	vasodilator-stimulated phosphoprotein	0.33	1.00E+00	**-2.60**	**1.39E-04**
OGT	O-linked N-acetylglucosamine (GlcNAc) transferase (UDP-N-acetylglucosamine:polypeptide-N-acetylglucosaminyl transferase)	-0.43	1.00E+00	**-2.58**	**2.20E-04**
JUND	jun D proto-oncogene	-0.84	1.00E+00	**-2.58**	**4.61E-04**
LOC648217	similar to ribosomal protein L37	0.33	1.00E+00	**-2.55**	**2.40E-06**
A_24_P844917	-	-1.49	2.38E-02	**-2.55**	**3.72E-03**
TBL1XR1	transducin (beta)-like 1X-linked receptor 1	0.43	1.00E+00	**-2.46**	**8.04E-05**
SMCHD1	structural maintenance of chromosomes flexible hinge domain containing 1	-1.31	7.46E-03	**-2.46**	**8.99E-04**
TOB2	transducer of ERBB2, 2	1.74	7.89E-05	**-2.46**	**4.89E-05**
VDAC1	voltage-dependent anion channel 1	0.07	1.00E+00	**-2.43**	**2.26E-03**
TBL1XR1	transducin (beta)-like 1X-linked receptor 1	-0.24	1.00E+00	**-2.43**	**1.59E-03**
ALKBH5	alkB, alkylation repair homolog 5 (E. coli)	0.66	1.00E+00	**-2.36**	**7.35E-05**
RBM25	RNA binding motif protein 25	0.00	1.00E+00	**-2.34**	**1.11E-03**
EVI5	ecotropic viral integration site 5	-0.22	1.00E+00	**-2.31**	**1.33E-03**
A_32_P31355	-	0.93	1.00E+00	**-2.30**	**1.09E-04**
A_24_P145911	-	-0.88	1.00E+00	**-2.28**	**1.82E-04**
KPNA4	karyopherin alpha 4 (importin alpha 3)	0.06	1.00E+00	**-2.26**	**7.45E-04**
MARCKS	myristoylated alanine-rich protein kinase C substrate	0.22	1.00E+00	**-2.24**	**1.37E-04**
KPNA6	karyopherin alpha 6 (importin alpha 7)	-0.46	1.00E+00	**-2.24**	**3.23E-03**
CEP170	centrosomal protein 170kDa	-0.18	1.00E+00	**-2.23**	**3.04E-03**
A_24_P936252	-	-1.51	1.04E-03	**-2.20**	**3.82E-04**
CPNE8	copine VIII	0.13	1.00E+00	**-2.20**	**6.63E-03**
A_32_P167577	-	0.53	1.00E+00	**-2.20**	**2.42E-05**
GAS5	growth arrest-specific 5	-0.74	1.00E+00	**-2.16**	**2.32E-03**
G3BP1	GTPase activating protein (SH3 domain) binding protein 1	0.94	1.00E+00	**-2.16**	**1.97E-04**
HIST1H4L	histone cluster 1, H4l	-0.85	1.00E+00	**-2.15**	**8.02E-03**
A_23_P92786	-	-0.56	1.00E+00	**-2.14**	**8.04E-05**
A_24_P910230	-	-0.30	1.00E+00	**-2.12**	**3.15E-03**
A_32_P175991	-	0.33	1.00E+00	**-2.11**	**1.05E-03**
HLTF	helicase-like transcription factor	-0.53	1.00E+00	**-2.09**	**2.82E-03**
EDEM3	ER degradation enhancer, mannosidase alpha-like 3	-0.04	1.00E+00	**-2.08**	**4.47E-03**
A_32_P70468	-	0.43	1.00E+00	**-2.08**	**4.36E-04**
RPS20	ribosomal protein S20	0.20	1.00E+00	**-2.07**	**1.11E-05**
CKAP4	cytoskeleton-associated protein 4	-1.42	1.88E-02	**-2.06**	**5.21E-03**
ERRFI1	ERBB receptor feedback inhibitor 1	-0.13	1.00E+00	**-2.06**	**3.17E-03**

Continued on next page

B.2 All genes found to be regulated in microarray analysis

Table B.2 – continued from previous page

Abbrev.	Name	Expr. tra	p-val.	Expr. tet	p-val.
SSR1	signal sequence receptor, alpha (translocon-associated protein alpha)	-0.20	1.00E+00	-2.06	3.10E-03
A_24_P500621	-	-1.47	4.13E-03	-2.06	1.45E-03
RPL39	ribosomal protein L39	0.30	1.00E+00	-2.01	1.15E-04
CYB5B	cytochrome b5 type B (outer mitochondrial membrane)	0.30	1.00E+00	2.05	1.10E-03
CCDC58	coiled-coil domain containing 58	-0.11	1.00E+00	2.06	2.81E-04
LOC401859	similar to peptidylprolyl isomerase A isoform 1	0.80	1.00E+00	2.10	2.62E-04
LOC390419	similar to peptidylprolyl isomerase A isoform 1	0.59	1.00E+00	2.10	9.03E-04
TM4SF1	transmembrane 4 L six family member 1	1.14	1.14E-03	2.17	1.68E-04
SPC24	SPC24, NDC80 kinetochore complex component, homolog (S. cerevisiae)	0.28	1.00E+00	2.19	1.50E-03
CNN3	calponin 3, acidic	0.62	1.00E+00	2.22	1.11E-03
A_24_P926450	-	-1.66	3.57E-04	2.27	1.78E-04
THOC4	THO complex 4	0.18	1.00E+00	2.28	1.77E-03
SAR1A	SAR1 gene homolog A (S. cerevisiae)	0.76	1.00E+00	2.31	8.62E-03
A_24_P901856	-	-1.64	1.20E-04	2.33	7.05E-05
A_24_P548415	-	-1.49	2.27E-04	2.46	7.78E-05
LOC728553	similar to ribosomal protein S8	0.50	1.00E+00	2.48	6.52E-03
THOC4	THO complex 4	0.31	1.00E+00	3.20	2.36E-03
A_32_P124773	-	-1.77	3.66E-03	3.83	2.98E-04

I want morebooks!

Buy your books fast and straightforward online - at one of the world's fastest growing online book stores! Environmentally sound due to Print-on-Demand technologies.

Buy your books online at
www.get-morebooks.com

Kaufen Sie Ihre Bücher schnell und unkompliziert online – auf einer der am schnellsten wachsenden Buchhandelsplattformen weltweit!
Dank Print-On-Demand umwelt- und ressourcenschonend produziert.

Bücher schneller online kaufen
www.morebooks.de

OmniScriptum Marketing DEU GmbH
Heinrich-Böcking-Str. 6-8
D - 66121 Saarbrücken
Telefax: +49 681 93 81 567-9

info@omniscriptum.com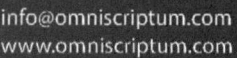
www.omniscriptum.com

Printed by Books on Demand GmbH, Norderstedt / Germany